Advances in Colorimetry

Edited by Ashis Kumar Samanta

Published in London, United Kingdom

Advances in Colorimetry
http://dx.doi.org/10.5772/intechopen.108083
Edited by Ashis Kumar Samanta

Contributors
Adwaita Konar, Ashis Kumar Samanta, Asis Mukhopadhyay, Ayan Pal, Chhaya Bhatt, Deepali Singhee, Dirk Labudde, Kakoli Dutta, Manikandan Krishna, Manish Rai, Nithya Sermugapandian, Pubalina Samanta, Ritwik Chakraborty, Rosalin Nath, Shree Krishna Shrestha, Sonia Karuppaiah, Sumithra Mohan, Tommy Bergmann, Yagya Kumari Shrestha, Yamini Dhanania

First published in London, United Kingdom, 2024 by IntechOpen
IntechOpen is the global imprint of INTECHOPEN LIMITED, registered in England and Wales, registration number: 11086078, 167-169 Great Portland Street, London, W1W 5PF, United Kingdom

British Library Cataloguing-in-Publication Data
A catalogue record for this book is available from the British Library

Additional hard and PDF copies can be obtained from orders@intechopen.com

Advances in Colorimetry
Edited by Ashis Kumar Samanta
p. cm.
Print ISBN 978-1-83769-183-8
Online ISBN 978-1-83769-184-5
eBook (PDF) ISBN 978-1-83769-185-2

For EU product safety concerns:
IN TECH d.o.o., Prolaz Marije Krucifikse Kozulić 3, 51000 Rijeka, Croatia,
info@intechopen.com or visit our website at intechopen.com.

Meet the editor

Dr. Ashis Kumar Samanta currently serves as a consultant in textile chemical processing. He brings extensive experience as a former professor and head of the Department of Jute and Fibre Technology at the University of Calcutta, India. Dr. Samanta has held leadership roles in various committees, notably chairing the TXD-07 committee on specialty Chemicals and dye stuff of the Bureau of Indian Standards (BIS) and the Board of Studies for the Bachelor of Fashion and Apparel Design (BFAD) program at the University of Calcutta. He is a valued member of WG-31 of the ISO/TC 38 Committee. With a commitment to advancing knowledge in textile science and engineering, Dr. Samanta serves as an associate editor for the AATCC Journal of Research and contributes to the AATCC's Publication Committee. He is also an esteemed editorial board member for prestigious journals such as the Indian Journal of Fibre & Textile Research (IJFTR) and the Journal of The Institution of Engineers (India) Series E. Dr. Samanta's contributions extend beyond academia, as evidenced by his involvement as a RAC and QRT advisory board member for the Indian Council of Agricultural Research-National Institute of Natural Fibre Engineering and Technology (ICAR-NINFET), and his role as an expert member of the Technical Research Advisory Committee for Khadi/Textiles at the Mahatma Gandhi Institute for Rural Industrialization (MGIRI), India. His significant impact on standards development is noteworthy, having contributed to the development of ten BIS Test Standards and five ISO Test Standards for the identification of natural dyes. Dr. Samanta has led 16 R&D projects and 2 Human Resource Development (HRD) training initiatives, showcasing his dedication to advancing the field. As a mentor, Dr. Samanta has guided fifteen Ph.D. students and has authored over 350 publications, including 150 research papers. His outstanding contributions have been recognized with prestigious awards such as the Dr. Triguna Charan Sen Award (1996 and 2014), the Bharat Nirman Award (2014), and an Eminent Engineer Award (2019).

Contents

Preface **IX**

Section 1
Advanced Methods and Tools for Color Measurement and Matching **1**

Chapter 1 **3**
Advanced Methods and Tools for Color Measuring and Matching: For Quality Check of Colored Products of Textiles and Apparel Industry
by Ashis Kumar Samanta

Section 2
Fundamentals and Applications of Colorimetric Analysis **43**

Chapter 2 **45**
Fundamentals of Colorimetry
by Yagya Kumari Shrestha and Shree Krishna Shrestha

Chapter 3 **67**
Application of UV-Visible Spectrophotometric Colour Analysis in Different Natural Product Identification
by Sonia Karuppaiah, Nithya Sermugapandian, Sumithra Mohan and Manikandan Krishna

Section 3
Applications of Colorimetric Analysis for Forensic and Defence Industry **83**

Chapter 4 **85**
Forensic Analysis of Bloodstain Color
by Tommy Bergmann and Dirk Labudde

Section 4
Applications of Colorimetric Analysis for Food Industry **109**

Chapter 5 **111**
Application of Colorimetry in Food Industries
by Kakoli Dutta and Rosalin Nath

Section 5
Applications of Colorimetric Analysis in Agriculture Enviornmental Medical and Pharmaceutical Industries **133**

Chapter 6 **135**
Colorimetric Determination of Pyraclostrobin Fungicide Using P-Amino-sulphonic Acid Coupling Reagent in Agricultural Soil/Environmental Samples by Spectrophotometric Analysis
by Chhaya Bhatt, Manish Kumar Rai and Joyce Rai

Section 6
Application of Colorimetric Analysis in Textile and Apparel Industry **155**

Chapter 7 **157**
Study on the Compatibility of a Binary Mixture of Babul Bark and Other Natural Dyes Used on Gallnut Pre-Mordanted Cotton Fabric for Developing Compound Shades
by Deepali Singhee, Yamini Dhanania and Ashis Kumar Samanta

Chapter 8 **183**
Colorimetric Characterisation and Process Standardisation for Application of Natural Dyes on Textiles: A Research Review
by Pubalina Samanta, Adwaita Konar and Asis Muhopadhyay

Chapter 9 **219**
Colorimetric Analysis to Study the Dyeing Behaviour of Jute Textiles Using Selective Synthetic and Natural Dyes
by Ritwik Chakraborty and Ayan Pal

Preface

Colorimetry is the science of quantitatively defining, measuring, and communicating a color of any solid, liquid, or semisolid material, instead of subjectively describing it, for evaluating and producing the same color or measuring total color differences and other color interaction parameters like K/S, L^*, a^*, and b^* values, assured by instrumental color measurements.

Color is determined by psychological and physical interactions and the chemical structure of materials influencing its applications. There are many books and monographs available on the basics of color science, colorimetry, color measurements, and so on. However, in today's digital age in which techniques of color evaluations and their applications have advanced greatly, there is a need for updated literature.

Color quantification via advanced techniques has become an essential part of color application for all industries. Initial simple colorimetric measurements have been replaced by spectrophotometric (absorption spectroscopy and reflectance spectroscopy) analysis of colors for different sectors. Moreover, for commercial exploitation, industries are conscious of the importance of color measurement, communications, and precision matching. As such, this book discusses both fundamental and advanced applications of colorimetric analysis for quality assurance of products in terms of color specifications and tolerances acceptable by consumers. The book also presents case studies on practical guidelines along with difficulties and solutions for colorimetric analysis.

The book includes nine chapters.

Chapter 1: "Advanced Methods and Tools for Color Measuring and Matching: For Quality Check of Colored Products of Textiles and Apparel Industry"

Chapter 2: "Fundamentals of Colorimetry"

Chapter 3: "Application of UV-Visible Spectrophotometric Colour Analysis in Different Natural Product Identification"

Chapter 4: "Forensic Analysis of Bloodstain Color"

Chapter 5: "Application of Colorimetry in Food Industries"

Chapter 6: "Colorimetric Determination of Pyraclostrobin Fungicide Using P-Aminosulphonic Acid Coupling Reagent in Agricultural Soil/Environmental Samples by Spectrophotometric Analysis"

Chapter 7: "Study on the Compatibility of a Binary Mixture of Babul Bark and Other Natural Dyes Used on Gallnut Pre-Mordanted Cotton Fabric for Developing Compound Shades"

Chapter 8: "Colorimetric Characterisation and Process Standardisation for Application of Natural Dyes on Textiles: A Research Review"

Chapter 9: "Colorimetric Analysis to Study Dyeing Behaviour of Jute Textiles Using Selective Synthetic and Natural Dyes"

Advances in Colorimetry is a useful compilation of comprehensive information on fundamental and advanced techniques in colorimetric measurements and its industrial applications. It is sure to be helpful for researchers, scientists, technologists, and industrial practitioners.

I wish to acknowledge the support of the staff at IntechOpen, without which this book would not have been completed. I am especially thankful to all the contributing authors for their hard work and repeated revision after a strict peer-review process.

Dr. Ashis Kumar Samanta
Professor,
Retired Professor and Ex-HOD,
Department of Jute and Fibre Technology,
Institute of Jute Technology,
University of Calcutta,
Kolkata, West Bengal, India

Section 1

Advanced Methods and Tools for Color Measurement and Matching

Chapter 1

Advanced Methods and Tools for Color Measuring and Matching: For Quality Check of Colored Products of Textiles and Apparel Industry

Ashis Kumar Samanta

Abstract

This introductory chapter for advances in colorimetry mainly addresses the basic principles of color quantification and the measurement of different known color parameters, color differences, dye uniformity, and tolerances for different parameters of color differences for assessing color matching efficacy. Few advances in the analysis of color-related issues for the textile and apparel industry have made this task easier with the aid of computer-aided color measuring and matching systems as indispensable tools. In addition, identification of colorant molecules on colored products by using UV–VIS spectrophotometry has become essential. Analysis of the quantifiable grade/rating of compatibility between two dyes for a pair of binary mixtures of colorants (dyes and pigments) to obtain a desirable shade has become a daily need in the production of dyed textiles and apparel products. Hence, in the present chapter, all such issues of advanced colorimetric analysis for color measurement and matching and quality checks of dyed/colored products of textiles and the apparel industry are discussed here along with the principles of working and design features of reflectance spectrophotometers. A few case studies with experimental data as examples are also elaborated here to understand these issues for better dissemination of knowledge in this field for industry colorists and researchers.

Keywords: color quantification and measurement, reflectance spectrophotometer, UV VIS-absorbance spectrophotometer, computer-aided color matching, dye compatibility, identification of color molecules

1. Introduction

Color is primarily a visual sensation in the human brain, producing relevant sensation in the retina of the human eye in the combined presence of a light source, a colored object and a perfect eye. The perception of color by the human brain is a color response related to the stimulation of the retina (comprising rod cells and cone cells) by the light spectrum reflected from the object after it absorbs a part of the light from the incident light spectrum of the light source. The retina of the human eye has two types of photosensitive cells: rod cells and cone cells. Rods are sensitive to dullness or brightness of light and are situated at the periphery of the eyeball. Cones are sensitive to color hues and are mainly situated in the forea region of the eye, and these cells are sensitive to color in three different wavelength bands—long, medium and short bands—which braineventually yield an output sensation interpreted as red, green and blue and/or proportional combinations of these three primary colors. The spectral sensitivity of the human eye corresponding to three primary/basic colors (red, green and blue). In the visible region of the human eye (400–700 mm), the color of the object is reflected by the complementary color wavelength or is sensed by the human eye after absorption of a certain wavelength by the coloring material (dye or pigment, etc.) from the source light spectrum. Hence, in a completely dark environment, we cannot see any object colored. The chemical constitution of the colorants (dye, pigment, etc.) is responsible for the absorption of part of the incident light energy. The human eye can observe and sense a color subjectively and can compare, but it cannot be compared quantitatively in terms of any numeric value for which human color perception differs from person to person for many reasons, such as (a) differences or changes in the light source spectrum or (b) differences in the observer's/sensitivity power, defective vision and eye fatigue. (c) Differences in direction and angle of vision. There are many other variable factors, such as adjacent color, background color, border color, sample size, surface texture and scattering, variation in observation angle and instrument setting.

According to Trotman [1], dyes are specific organic molecules that can absorb a certain part of the light selectively and can reflect the other part of the light in the visible zone. Since then, many investigations have reported on which part of the dye molecules of natural/synthetic dyes/pigments are responsible for the visible color of that object, with a predominant hue at a specific wavelength.

According to Shah and Gandhi [2], color is defined as a psycho-chemical phenomenon that involves a psychological stimulus to the human eye. Therefore, to a chemist, it is a chemical compound with conjugations; to a physicist, it is a multiple physical phenomenon occurring simultaneously, such as the reflectance, absorption and scattering of light waves; to a physio-logist, it is a sensory measurable response of electrical signals produced to nerves of human eyes; and to a psycho-logist, it is a sensation of mind in response to reflected color waves. To artists and other common people, it is a sensory stimulant in the human brain that is used as an observer.

Ordinarily, for many industrial processes and products, it is difficult to use the accuracy of the appearance of color to the human eye because it is difficult to differentiate one color from another by only the red/blue/green responses of the human eye. According to an earlier report by Samanta [3] on the use of computer-aided color measurement and matching for textiles, apparels, paint, leather, pharmaceuticals, polymers/plastics, paper and food, etc., it is necessary to work hard to prepare a color database for checking and maintaining color matching quality, as the color parameters of these products, along with other physical or chemical properties, are important product specifications. Earlier, color specifications were assessed by visual methods of human eye perception, but currently, assessments of color strength produced for particular hues and color differences for color matching from given or desired standard shades have started to be performed instrumentally by computer-aided color measuring systems. Hence, computer-aided color match prediction has also gradually started in all these industries for precision matching.

There are many disputes and misconceptions [2, 4] regarding match/nonmatch decisions in color matching, and such decisions are assessed via visual human eye procedures. Therefore, instrumental precision measurement techniques for practical color measurement and for checking color matching results in terms of quantifying color differences on red, blue, green, yellow and dark or light scales should be considered better options than human-perceived color (assessed by visual methods).

In any industry [2–4], the desired matched color is obtained by the application of a mixture of two, three or four dyes/pigments. In a few industries, an experienced senior dyer, called a 'shade bank', maintains a record of hard copy shades produced with a small piece of attached sample for color matching of different substrates from his day-to-day experience. Therefore, for the next sample for color matching, Sr.

An experienced dye master first selects one of the color recipes from his or her shade bank that is very nearer or very closer to the color of the standard sample given for matching. Then, color is added or subtracted, and nearer or nearest matching is performed via trial and error of actual dyeing. However, the present computer-aided color match instrument can be obtained from the saved color database of that class of dyes. This instrumental system can predict much closer matches with expected color parameters and color strength to produce within limits of set tolerances within a few minutes. Even if exact color matching is not obtained in the first predicted formulations, it is possible to add or subtract a certain amount of color in the predicted formulation either automatically via menu or via a computer-aided color matching system to obtain the exact match or manual addition.

Due to the increase in the variety of natural and manmade fibers and their use in multifiber blended textiles and the increased availability of synthetic dyes and pigments available on the market, an increasing number of companies have made them available. Therefore, special care of the nature and type of fibers and the type of dyes used are also important for determining color recipe formulations using computer-aided color measuring and matching systems. The color matching formulation should also be the most economical and least metameric, reproducible, most uniformly dyed, and have moderate to good washing and lightfastness properties in the matched product as a better quality colored textile product, which is not easy to achieve, without the use of computer-aided color measuring and matching tools [3, 5].

This new technology of using computer-aided color measuring and matching tools enables expert computer-aided dye masters to easily predict the number of color matching formulations/recipes with multiple options to select the most appropriate least expensive and least metameric recipe [2–5] for color matching depending on the current price of the dyes and degree of precision in matching by measuring color differences (in terms of the DE, DL, Da and Db values and the Labd metamerism index value).

The computer-assisted color-measuring and color-matching system, which is applicable for any textile industry, apparel industry or other industry, is therefore considered a powerful tool for improving the quality of precision color matching of products and provides a great opportunity to maintain less dye inventory, reduce the cost of production and reduce the time required for predicting and producing a good color matching recipe.

The colored molecules (dyes and pigments) can also sometimes be identified from colored textiles and other products to assess the nature of the color molecules used and their properties, eco-friendliness, etc. Colored molecules can be identified from colored textiles by subjecting them to the extraction of those colored ingredients (dyes or pigments) by water or any other solvent extraction process to extract colorants from colored textiles and apparel products using suitable solvents under a known method of such extraction; then, the extract can be subjected to UV–VIS spectroscopic analysis to obtain wavelengths vs. optical density/absorbance curves as fingerprints of particular color ingredients/molecules present there. This may also be confirmed by chromatography, HPLC, NMR, LC–MS and FTIR analysis. However, after dyeing textiles with a mixture of colors to obtain a particular compound shade, it is more important in the industry to measure color variation for variations in dyeing conditions, dye and mordant (in the case of natural dyeing) concentrations and % application/proportions of different dyes to obtain a perfect match of color/shade within certain prescribed and acceptable limits of tolerances of different color parameters. For compound shades, another parameter known as colorant/dye compatibility has also become a very important factor for producing a shades of precision match.

Davidson et al. [6] first introduced an analogy for using a computing device for predicting a color recipe from a preprepared stock of color database used for color matching of wall paint initially and gradually, which was successfully applied to textiles and apparel products.

At present [3, 4], almost every dye house in the textile industry uses computer-aided color measuring and matching systems. To generate/predict the color matching recipe from the preprepared color database, we checked the instrumental color differences to check the quality of the precision matching results and to carry out batch correction to achieve a better match if the predicted recipe did not work in one shot. A computer-aided color measuring and matching system therefore involves three basic assemblies, viz.:

1. Color measuring unit/instrument: A reflectance spectrophotometer, which measures the reflectance values of any object initially and then expresses the color values in numerical form quantitatively for any desirable color parameters, which are easy to process/display on a computer to obtain a meaningful color parameter.

2. Color matching software: Dedicated software is used for precision color measurement and prediction of color matching formulation with the aid of a suitable reflectance spectrophotometer as an instrumental aid for subsequent color measurement and match prediction sequences by expressing the color-related measured/stock data converted to a desirable color parameter that is relevant for obtaining direct results/matching formulations, as per the requirements of the colorist.

3. Computer hardware: A PC with high resolution and high storage capacity and a quick data processing system to operate, detect and display the resulting data and curves.

2. Basic principles of the use of computerized color measuring and matching tools

Precision color matching [2–4] for textile substrates is mainly based on the Kubelka munk function [2–6], Park and Stearn's [7] algorithms and Allen's [8] matrix for generating closer recipes via computer-aided color matching formulation [9] programs using a reflectance spectrophotometer, a suitable color database for specific substrates and specific dye classes, and dedicated software for calculations of required color parameters and match prediction formulation programs processed on a can computer attached to a spectrophotometer.

Multiplication of the variability of the dyeing process to achieve accurate color measurements via computer-aided color matching systems requires preoptimization of the dyeing process variables and minimization of all other processing and measurement errors in the preparation of color databases and in the generation of color matching formulations [3, 9] by assumptions, approximations, simplifications and iterations, etc., considering that color matching of any product is an iterative trial and error process. Manual matching of color by expert/skilled colorists/divers can never achieve that level of precision in the color matching of any product as a computer-aided system can. Therefore, manual color matching is not easily accepted by customers, as manual color matching shows wide variations or differences in color variation, as human eye perception varies from person to person. For instrumental matching, variation from illuminate to illuminate or instrument to instrument is also present, but this variation may be minimized by choosing the least metameric match. Thus, for instrumental color matching vs. manual color matching, the former (instrumental color matching) has many advantages over manual processing.

The basic process of computer-aided color matching [2–5, 9] for predicting color matching formulations for the textile and apparel industry involves the following five steps:

1. Preparation of the color database and its storage (color type or dye classwise, companywise and substratewise)

2. Determination of the optical constants or K/S values of all database samples and standard samples

3. Prefixing allowable color difference tolerances for the prediction of acceptable formulations under standard illuminants

4. Comparison of the least-cost and least-metalistic color matching formulations after an actual dyeing trial

5. Batch correction for any deviation between the predicted formulation and actual dyeing trial results.

The optical constants of the Kubelka Munk function [2–4], i.e., K/S values (surface color strength), are initially measured after a textile substrate is dyed with each dye at a multilevel concentration to obtain a linear relationship between dye concentrations and K/S plots for each dye after dyeing a specific textile substrate with different concentrations of colorants of a specific class of dyes from the same company at 8–10 different levels and with 1 sample as a control undyed sample as the substrate. All calibration of the color database was performed by using optimized dyeing process parameters for uniform dyeing results. Accurate color database preparation is the basic need for generating precision color matching formulations within specific tolerances of multilevel RBGY scales of color differences. After checking the accuracy of the prepared color database, these data were stored in a computer-aided color matching library system in a separate file for future use. The accuracy of the color database for each colorant should be checked by linearity [3, 5] in each case by plotting the dye concentrations vs. K/S values, and if it is not linear, either redyeing or approximation/deletion of some particular points or linear curve fitting should be performed appropriately. Once the color database is ready, the generation of color match formulations can be performed by following the iterative process of the color match prediction algorithm following the flow chart given below (as shown in **Figure 1**), where these process steps remain in built-in dedicated software presented in the computer-aided color matching system):

For further correction of shades, the reformulation of batch correction mathematics for computing the incremental value of the incremental concentration of one or two colors for different dyes may require further iteration at that stage, which may be represented mathematically as follows:

$$DC_1 = -\frac{\partial C_1}{\partial X}DX + \frac{\partial C_1}{\partial Y}DY + \frac{\partial C_1}{\partial Z}DZ \quad (1)$$

$$DC_2 = -\frac{\partial C_2}{\partial X}DX + \frac{\partial C_2}{\partial Y}DY + \frac{\partial C_2}{\partial Z}DZ \quad (2)$$

$$DC_3 = -\frac{\partial C_3}{\partial X}DX + \frac{\partial C_3}{\partial Y}DY + \frac{\partial C_3}{\partial Z}DZ \quad (3)$$

The dye concentrations of the colorants (C_1, C_2, C_3 or Cn) used for obtaining a match recipe and X, Y and Z are the tristimulus values of the colored/dyed samples, and the DX, DY and DZ values are the differences in the Tristimulaus values between the produced samples and standard samples or differences between new sample

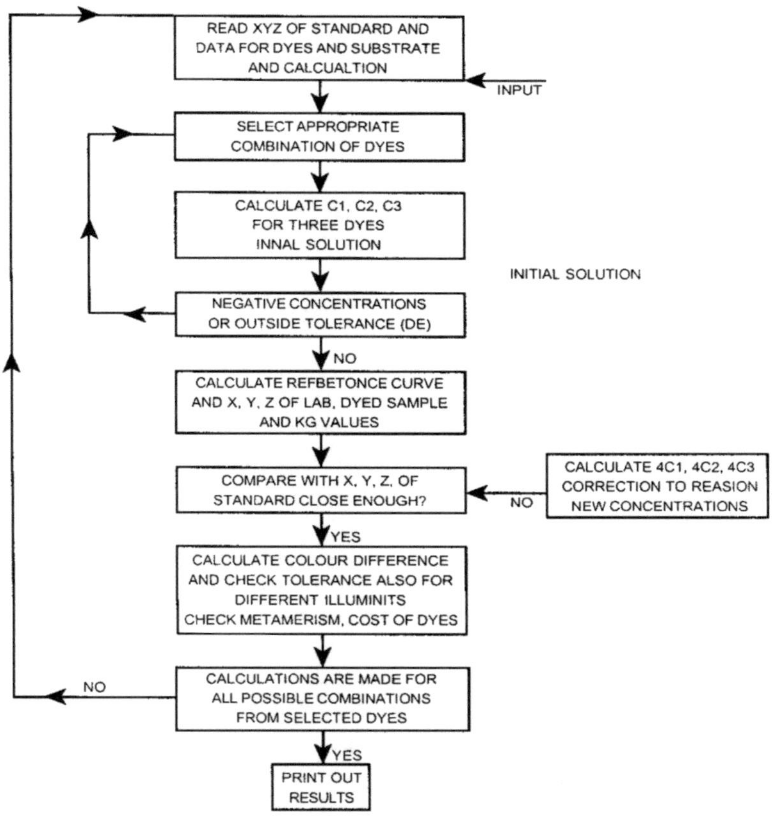

Figure 1.
Flowchart of the computer-aided color matching algorithm for the prediction of color match formulations.

batches and produced samples dyed with the predicted color matching formulation; further batch correction is needed for precision matching.

To generate a recipe formulation for the color match prediction process, measurements of the color of standard shade must first be performed to match a specific substrate, and then, dyes can be selected according to the dye inventory/color database type available for use by assigning or fixing the required tolerance limits [2, 3, 5, 6] for color difference values. Then, the computer color matching system starts determining/predicting possible match formulations from specific dye class database sources available/stored in the system after actual dyeing. The color values, in terms of the tristimulus values of the standard sample to match, are computed under a specific light D65 source with the predicted match formulation. By comparing them with their color data and optical constants, for example, for a predicted formulation from a specific class of dyes and its stored color data bank, the unknown concentration of a specific dye mixture required for matching with a particular color standard can be determined with the help of software.

If the predicted recipe falls practically after actual dyeing, it needs to be corrected by the batch correction program. Matching of shades for the blended textile fabric also essentially needs to be carried out similarly, but the data banks with different classes or types of dyes (according to the fiber type in the blends) applied to both the components of the blended textiles are stored as specific color databases for blended components separately for use in match prediction on blended textiles. Dan and Randall [10] clarified the need for uniform dyeing (with less than 5% CV for K/S values at 10 different points for colored textiles) and explained that the preparation of reliable calibration dyeing (with checking of linearity of dye concentrations vs. K/S values in calibrated dyeing series) for the preparation of a color matching database and its successful storage (type of dye classwise, manufacturer/companywise, etc.) is the first and foremost very important step for the generation/development of a computer-aided color match prediction for any substrate within a range of acceptable color differences [11].

The solutions of the above color matching equations for computer-aided color match prediction (CCMP) usually follow those of the Kubelka munk function [2–6], Park and Stearn's [7] algorithmic equation and Allen's [8] color matrix for computer-aided color matching formulation [9] program. An isomeric match is a match under all illuminations, while a two-color sample that appears as a match under one illuminate but does not match under any other illuminate is termed illuminate metamerism, and that type of match is called a metameric match. Our objective is to find the minimum metameric or least metameric match prediction, and formulation sorting is based on the determination of the general metamerism index or the illuminance of the metamerism index to maintain the least metamerism. Some improvements in computer-aided color matching formulations based on minimum color difference tolerances, fewer or least metameric effects, etc., are needed [12, 13].

However, due to practical variations in the dyeing conditions, machinery settings and human interventions of dyeing operators, there may still be some differences in the actual color yield of the products dyed by the predicted formulations; however, some corrections, called batch corrections, are needed to achieve/improve the desired precision of the color matching quality of the produced sample against the color of the given standard dyed product [14]. Recently, a new test method for analyzing dye compatibility via relative quantitative rating/grades for the use of a mixture of colors to obtain uncommon compound shades on textiles has been established that is applicable for both natural dyes [15–17] and synthetic dyes [18], the details of which are available from the current relevant literature.

The effects of dyeing process variables on the color yield and color fastness properties of cotton khadi fabric dyed with deoiled Redsandal wood waste [19] were analyzed via colorimetric analysis after cotton fabric was dyed with red-sandal wood waste as a natural dye.

Improvements in the color fastness of naturally dyed textiles and their antimicrobial properties have been reported in recent literature via detailed colorimetric analysis [20]. The application of computer-aided color measuring and matching instruments for the textile and apparel dyeing industry has been comprehensively described by P Samanta [21] due to their advantages and disadvantages.

Android phone-based color measurement and color data analysis, i.e., digital imaging techniques for color parameter tests and color fastness assessment with DigiEye software to achieve maximum accuracy, are of interest to color scientists and industry dyeing managers. The Clarion color fastness app has already been developed in this direction by Kuraray, Japan.

IS standards for the identification and determination of the purity of natural indogo have recently been standardized by BIS [22], which is also discussed here as a case study, as an important application of colorimetric analysis in natural dyed textiles.

The basics of the quantitative analysis of colorimetric parameters for colored textile and apparel products are therefore depicted below to clarify the subject clearly for proper analysis.

2.1 Principles of color quantification and measurements

Color can be measured quantifiably and absolutely by its three coordinates known as tristimulus values, i.e., X, Y and Z, or by total reflectance ($R\lambda_{max}$) values or its derivative formula function, such as K/S values. The CIE tristimulus value [2–5] of a colored substrate may be defined as

$$X = \Sigma P_\lambda X_\lambda R_\lambda \tag{4}$$

$$Y = \Sigma P_\lambda Y_\lambda R_\lambda \tag{5}$$

$$Z = \Sigma P_\lambda Z_\lambda R_\lambda \tag{6}$$

where P_λ = the spectral power distribution of the standard source.

R_λ = Spectral reflectance of the substrate.

$X_{\lambda.}$ $Y_{\lambda.}$ $Z\lambda$ = color factor of the standard observer for the red, blue and green sensations of the human eye.

To describe color in a two-dimensional plot called the CIE color space diagram, the CIE is defined by the following chromaticity coordinates (x, y, z) to measure color and color differences:

where

$$x = \frac{X}{X+Y+Z} \tag{7}$$

$$y = \frac{Y}{X+Y+Z} \tag{8}$$

and

$$z = \frac{Z}{X+Y+Z} \tag{9}$$

and x + y + z = 1 from which the saturation can be determined, and from the two third can be determined.

A color match [2–5] means that the color of the produced sample (SL) is the color of the standard (SD), i.e., $(X_{SL}, Y_{SL}, Z_{SL}) = (X_{SD}, Y_{SD}, Z_{SD})$, while X, Y & Z are the tristimulus values of the sample (SL) and standard (SD) or $(\text{reflectance})_{SL}$ (400 to700 nm) = $(\text{reflectance})_{SD}$ (400 to 700 nm) or $(K/S)_{SL} = (K/S)_{SD}$, while $K/S = \alpha\, C_D$, where K/S is the Kubelka Munk function called color strength, which is the coefficient of absorption divided by the coefficient of scattering, and C_D is the concentration of dye/colorants used.

The CIE theory of color–Tristimulus values (X, Y and Z) and CIE color communication parameters updated in 1976, i.e., DE*, DC*, DH*, L*, a* and b* values, etc., are well described by Samanta [3] with descriptions and figures in a book chapter of the Intech open published book on colorimetry.

For a mixture of colorants used to match a compound shade, the following three equations are solved as a function of dye concentrations of the colorants (C_1, C_2, C_3 or Cn), and tristimulus values can be measured through reflectance measurements [2–5].

$$f(C_1, C_2, C_3, \text{or}\, Cn) = X, f(C_1, C_2, C_3, \text{or}\, Cn) = Y \text{and} f(C_1, C_2, C_3, \text{or}\, Cn) = Z \quad (10)$$

where X, Y, and Z are the tristimulus values of the samples to be matched and C_1, C_2, and C_3 or C_n are the concentrations of dyes or color pigments needed. In practice, the reflectance values at 400 to 700 nm are measured from solid-colored textile surfaces, and those reflectance data are processed for K/S value generation for ultimate matching. The reflectance vs. concentration of dye is nonlinear and nonadditive, so the reflectance vs. dye concentration data cannot be directly used as basic data for color match prediction; rather, plots of dye concentration vs. K/S values are linear, and the K/S values are additive. The K/S data can be used for computer-aided color match prediction. The empirical relationship of the Kubelka Munk function [2–5] (i.e., K/S values) varies linearly with dye concentration.

$$\begin{aligned} K/S &= \frac{\text{Co-efficient of absorption}}{\text{Co} - \text{efficient of scattering}} \\ &= \frac{(1 - R_{\lambda max})^2}{2R_{\lambda max}} = \alpha\, C_D \end{aligned} \quad (11)$$

where R is the reflectance value at a chosen user wavelength or at λ_{max} (at the maximum absorbance wavelength).

Thus, a higher K/S indicates a higher absorption for dyes or a mixture of dyes, indicating a higher absorption value or a higher dye uptake.

Thus, $K/S\ \alpha\ C_D = \alpha\ C_D$, where α is a constant.

As K/S is also additive in nature, for mixtures of colorants used to dye textile fabrics for compound shades, the total K/S will be or can be determined by the following additive relationship [2–5]:

$$(K/S)_{\text{total after dyed with Mixture of Dyes}} = (K/S)_{\text{undyed subs}} + (K/S)_{D1} + (K/S)_{D2} + (K/S)_{D3} \dots \dots + (K/S)_{Dn} \quad (12)$$

$$\text{Or}(K/S)_{\text{dyed substrate}} = (K/S\,)_{\text{undyed subs}} + \alpha_1 C1d_1 + \alpha_2 C_2 d_2 + \alpha_3 C_3 d_3 \dots \dots \dots \dots \dots + \alpha_n Cnd_n \quad (13)$$

For dyes on textiles, it is assumed that dyes do not contribute to scattering, and if the substrate is not changed, the scattering of the substrate also remains constant, while K is the coefficient of absorption, i.e., the sum of dye stuff absorption and substrate absorption (as the substrate is fixed, there are no changes in scattering due to the substrate). Therefore, K/S directly varies with the concentration of dyes, and scattering is independent of dye concentration (which is not the case for pigments in paint).

For textiles, for a particular sample (for which the material, yarn and fabric construction and surface finish remain unaltered), the scattering remains constant. Therefore, in textiles, this theory is called the single constant theory. $K/S = \alpha C_D$.

Finally, the reflectance vs. dye concentration is not linear, and hence, it is difficult to interpolate or curve fit to predict the achievable color strength from any of the reflectance or tristimulus values of dyes.

The K/S vs. dye concentration is linear; hence, it can be interpolated to any desired dye concentration and thus can be safely used [2–5] in computer-aided match prediction software to predict color matching formulations for mixtures of two or three dyes or more.

While

$$L^* = 116(Y/Y_o)^{1/3} - 16 \Delta L^* = L*_1 - L*_2 \tag{14}$$

$$a^* = 500\left[(X/X_o)^{1/3} - (Y/Y_o)^{1/3}\right] \Delta a^* = a*_i - a*_2 \tag{15}$$

$$b^* = 200[Y/Y_o)^{1/3}\Big] - (Z/Zo)^{1/3}\Big] \Delta b^* = b*_1 - b*_2 \tag{16}$$

The chroma (psychometric chroma) values in the Cl $_{104}$ color space were calculated as follows:

$$C_{ab}{}^* = \left(a*^2 + b*^2\right), \Delta C^* = C*_{1(ab)} - C*_{2(ab)} \tag{17}$$

where $C^*_{1(ab)}$ and $C^*_{2(ab)}$ are the chroma values for the standard sample and produced sample, respectively.

CIE 1976 metric Hue-Difference ($\Delta H_{ab} = [(\Delta E_{ab}{}^*)^2 - (\Delta L^*)^2 - (C_{ab}{}^*)^2)]^{1/2}$.

An isomeric match, i.e., a match under all illuminants and when two colored samples match under one illuminant but do not match under any other illuminant, is termed a metameric match [2–5]. Least metameric match prediction and sorting are based on the calculation of the general metamerism index [2, 3], as given below:

$$\text{General metamerism Index} = \frac{\sum (DR\bar{x})^2}{X^2} + \frac{\sum (DR\bar{y})^2}{Y^2} + \frac{\sum (DR\bar{z})^2}{Z^2} \tag{18}$$

where DR = Difference in reflectance between a pair of metamers;

x, y, z = CIE standard observer color functions;

X, Y, Z = the CIE tristimulus value normally taken for illuminant C, which is the average of two specimens.

Additionally, the CIE LAB, i.e., the LABD metamerism index [2–5], is calculated from the following CIE Lab equations [2, 3]:

$$MI_{(LABD)} = \left[(DL_1{}^* - DL_2{}^*)^2 + \left(Da_1{}^* - (Da_2{}^*)^2 + \left(Db_1{}^* - (Db_2{}^*)^2\right]^{½} \tag{19}$$

where DL_1^*, Da_1^*, and Db_1^* are the Delta CIE Lab* -1976 color coordinates [2, 3] between Standard and Sample for the first illuminate and DL_2^*, Da_2^*, and Db_2^* are the Delta CIE Lab* color coordinates between Standard and Sample for the second illuminate.

2.2 Methods for predicting color matching formulations

The prediction of color matching formulations is based on the use of binary or ternary mixtures of different colorants to obtain the resulting compound shade. Effective coloration with a mixture of dyes depends on the compatibility between dye and dye, i.e., the rate of color build-up when dyed under a specific optimized set of dyeing conditions. Therefore, before predicting color matching formulations, determining dye compatibility between any two dyes/pigments is another important parameter. Along with visual assessment, the rate of dyeing was assessed, and the coefficient of the dye diffusion test was assessed to determine the rate of progressive color build-up of both individual dyes in a binary mixture of two dyes when dyed under optimized dyeing conditions. The latter is a very popular conventional method to test dye compatibility, i.e., to determine the rate of color build-up of two individual dyes when the substrate is dyed with a binary mixture of dyes together, and the textile substrate under reference is to be dyed under two different sets of dyeing conditions for testing dye compatibility (in one set by varying the profile of dye concentration while keeping other dyeing conditions fixed and in the second set of dyeing conditions is to be done with variations in the dyeing time and dyeing temperature, keeping the dye concentration and other dyeing parameters fixed). After these two sets of dyeing under different conditions, the rate of increasing chroma/hue for gradual color build-up with increasing dyeing time/temperature and with increasing dye concentration were checked by comparing the pattern of color built-up by plots of K/S Vs DL and DC vs. DL in this conventional method. However, this conventional method of dye compatibility testing has many limitations, such as being time consuming, requiring high analytical skill and being cumbersome to follow and not quantifiable in terms of the degree of compatibility. Hence, a newer and simple advanced method of dye compatibility analysis between two dyes/pigments, which is simpler, quicker and easier to apply, has been established by Samanta and Agarwal [15, 16].

The second important issue before computer-aided color match prediction for any textile substrate is the preparation and storage of an accurate color database to store the dyeing results of calibrated dyeing samples on the same textile substrate with a specific class of 9–10 dyes of different colors after checking the accuracy of the dyeing results by checking the linearity of the dye concentration vs. K/S values for each colorant and after checking the dye uniformity (less than 5% CV of the K/S values).

Therefore, after measuring the color of the dyed samples under commonly used illuminants such as D65, the reflectance, K/S values and L*, a*, and b* values resulting from the calibrated dye were recorded and stored in a color matching computer system.

The reflectance spectrophotometer measures reflectance values at different wavelengths from 400 to 700 nm at 10 nm intervals, and these data are used by computer-aided color matching software to calculate the DL*, Da*, Db* and DE* scale of color difference values for the newly dyed/produced sample compared to the standard colored fabric to match. Currently, computer-aided data for each class/type of dye and each color of that specific class of dye stored in the system are used by computer-aided color match prediction software based on the linearity and additive nature of the relationship between dye concentrations and the K/S value for the prediction of a particular color match using different binary or ternary mixtures of dyes, i.e., using the basis of the following CIE equations to calculate the total K/S value of the dyed sample to match the K/S value of the standard sample, as already shown above and described below.

$$(K/S)_{\text{dyed with mixture of dyes}} = C_0(K/S)_{\text{un dyed substrate}} + C_1(K/S)_{D1} + C_2(K/S)_{D2} + C_3(K/S)_{D3} \tag{20}$$

$$\text{Or } (K/S)_{\text{dyed substrate}} = C_0(K/S)_{\text{un dyed substrate}} + C_1(K/S)_{D1} + C_2((K/S)_{D2} + C_3(K/S)_{D3} \tag{21}$$

Thus, this method is also based on the gradual increase in the color yield from the mixture of different dyes used in the mixture, the matching of the total K/S values between the standard sample to match and the total color strength (K/S) value of the newly produced sample, which are the determining points of match predicting formulations showing the amount required for C1 concentrations Dye-D1 and C2 concentrations of Dye-D2 and C3 concentrations of Dye-D3. Finally, this method provides the required information about the computer-aided match prediction formulation (S) in more than one or as many as nos of different formulations are possible with similar dyes/colorants having closely similar hues, thereby proving/making a computer-aided color match prediction system, as a more advanced tool, for better quality and precision color, match prediction for more practically useful purposes of industrial dyers/colorizts.

Color differences in terms of DL (lighter or darker), Da (redder or greener) and Db (bluer and yellower) values showing less or more darker or lighter, redder or greener and bluer and yellower indicate the need for batch corrections to eliminate minor differences in any color difference notation to meet the desirable or preset tolerances of DL, Da, Db and DE values. However, in practice, dyers feel that it is difficult to control any of the color differences among DL, Da and Db without affecting other tolerances.

Hence, another newer index of color differences, known as the color difference index (CDI) [15, 16], was recently established to obtain an easy understanding of overall color differences and the dispersion of color on the fabric surface after dyeing for different dyeing variables/conditions via the following equation (considering the magnitudes of the relevant ΔE, ΔH, ΔC, and MI values (ignoring their sign and direction):

$$\text{Colour Difference Index(CDI)} = \frac{\Delta E \times \Delta H}{\Delta C \times MI} \tag{22}$$

A color difference index (CDI) less than 5 is considered acceptable for preparing a color matching database because this difference is within the tolerance limits of color differentiation. However, the higher the color difference index (CDI) value is, the greater the nonuniformity in dyeing, which has a wide variation in color strength, according to a comparison of one dyed textile sample with the other, and that match is not acceptable.

This CDI index is also currently used in the relative compatibility rating (RCR) method for determining compatibility between any two dyes for use in binary mixtures of two natural or synthetic dyes, where the differences in the maximum and minimum CDI values are used as a reference for determining the rate of color build-up in the relative compatibility method after dyeing with different proportions of binary mixtures of those two dyes to determine the numeric rating of compatibility between those two dyes under test, from 0 to 5 ratings of compatibility by the said RCR method [15, 16].

Additionally, another newer index, the color matching index", is defined by the following empirical relationship, depending on the compatibility of the dyes used in a mixture for compound shades for checking the appropriateness of predicted color matching formulations using an appropriate color database after maintaining a linear relationship between the K/S and dye concentrations in computer-aided color match prediction predictions of color data. The compatibility between any two dyes used in a binary mixture (by dyeing under 2 different sets for checking their color build-up rate), (i) by varying the overall dye concentration (taking 50:50 of each dye) in one set while keeping the dyeing conditions fixed and (ii) in another set by varying the dyeing temperature and time profile gradually keeping the dye concentration fixed and then plotting K/S Vs DL and DC vs. DL to check their color build-up rate, which may mathematically be expressed as follows (for the above two sets of dyeing results for the progressive build-up of color in compound shade by a binary pair of any two dyes):

$$\text{Dye Compatibility Index}(\text{between any two dyes used in binary mixture}) = \frac{K/S/\text{DL}}{\text{DC}/\text{DL}}\,\text{i.e}\,\frac{K/S}{\text{DC}} \tag{23}$$

When the DCI is higher, the dye pair is less compatible, and when the DCI is lower, the dye pair is more compatible. However, this newer index has not yet been widely tested and needs to be tested in different cases to understand/prove its efficacy compared to other methods of determining dye compatibility.

Another newer index, the color matching index (CMI), can also be expressed by the following empirical relationship (irrespective of their sign and direction, ignoring any negative or positive sign) by comparing the color data of the standard sample and produced matching sample, as given below:

$$\text{Color matching Index}(\text{CMI}) = \frac{K/S \times \text{DC}}{\text{DL} \times \text{DL} \times 100} \tag{24}$$

Additionally, CMI values above 5, in any case, are not considered acceptable for compatible dyes to match any standard shade and can be stored in computer-aided color matching databases. However, this CMI value must be widely used by textile industry colorists to understand its true significance in textile color matching.

2.3 Color measuring instruments

The human eye cannot quantify color quantitatively in terms of some useful colorimetric functional terms/data. The comparison perceived by the human eye is subjective only. Moreover, assessments of color vary from person to person in terms of the viewing angle, eye sensitivity, vision defectiveness, eye fatigue, etc. Additionally, color comparisons by human eyes differ with respect to the light source, which is affected by the background/boarder and adjustment, and cannot be used to determine exact reproducibility in repeated assessments in addition to the above eye limitations.

Therefore, there is a need for an instrumental repeatable measurement technique for a color surface or solution in quantitative terms that can provide reproducible results at the saved preset condition of measurement, where the conditions of measurement and detection can be controlled to generate quantitative colorimetric data suitable for comparison in quantitative terms, in addition to facilitating color communication in specified quantitative terms after instrumental measurements.

Fundamentally, there are two types of color measuring instruments—(a) a tristimulus colorimeter and (b) two types of UV–VIS color spectrophotometer—one for measuring the color of a colored solution and the other for measuring the color of colored solid samples.

2.3.1 Tristimulus colorimeter for color measurement

A Tristimulus colorimeter is an instrument used to obtain color response functions that are directly proportional to those of the CIE standard colorimetric values of red, green and blue primaries in terms of tristimulus values (X, Y, and Z). In this instrument, the radiant power from the light source is incident to the colored objects. This reflected radiant power is measured through three tristimulus filters, and the filtered responses fall on the photodetector to determine the spectral response proportion to the corresponding tristimulus values (X, Y, and Z) of the object under a standard light source. These raw data can be transferred to a microprocessor for the comparison of the absolute CIE trust values (X, Y and Z). Colorimetric measurements are thus very easy and quicker to perform with this low-cost colorimeter. However, these instruments can measure only tristimum values and are less accurate; hence, they cannot provide an overall assessment of all types of colorimetric parameters.

2.3.2 A UV–VIS spectrophotometer was used to measure the color of the solution/extracts or solid samples

Hence, for more accurate color measurements, two types of UV–VIS colorimetric spectrophotometers have been designed. There are two types of UV VIS spectrophotometers used for color measurements: UV–VIS absorption spectrophotometers and UV–VIS reflectance spectrophotometers [2, 3]. Thus, (i) an absorbance spectrophotometer can measure the color in liquid or extracted solutions (a) in terms of the absorbance (Abs) of the liquid color sample, and (ii) a reflectance spectrophotometer can measure the color at the surface of the colored solid sample in terms of (b) the surface reflectance (R) of the surface of the colored solid sample at the chosen/preset wavelength or the absorbance maxima wavelength in both the visible region and the UV region for the measurement of the color values of any liquid or solid sample, respectively.

For computer-aided color measurements and matching systems, color measurements, quality control checks of the color produced, match predictions and other arrangements of analysis of color data in different modes are required to judge the efficacy of computer-aided color measuring and matching systems using a reflectance spectrophotometer with a computer and the associated software, which is briefly explained here.

2.3.2.1 Principles of using a reflectance spectrophotometer as a color measuring instrument

The use of reflectance spectrophotometer color measurement techniques has become very popular and widely accepted in the textile, leather and paint industries because of the advantages of linking computer-based operation technology with color measuring spectrophotometric instruments for storing, analyzing and comparing old and new color data at finger tips, particularly with the advent and development of friendly color measuring and match prediction software.

Reflectance spectrophotometers [3] are used to measure the reflectance value or reflected radiant power in terms of red, blue and green responses, which are the same as those of a standard sample at any desired prefixed wavelength or at a predetermined maximum absorbance point. The reflectance spectrometer for color measurement consists of a light source. The emitted light falls incident on the surface of the object at a definite angle or as diffused light on to the object, while the resultant reflected light passes through a mono-chro-meter (of chosen wavelength) or different monochrometers for a measurement in order to arrange a wavelength at a predetermined wavelength value. The mono-chro-meter disperses the incident radiant energy of source light spectrally and transmits it via a narrow band of wavelength through the exit slit. The detector carries a detection system that receives the spectral radiant power reflected from the object and from the standard sample to generate a ratio of spectral signal that is transmitted to the computer for analysis and displayed in different colorimetric terms according to the color measurement software design. With the fundamental reflectance data, it is possible to compute colorimetric data of various types as needed for different applications.

2.3.2.2 Design features of a good color-measuring reflectance spectrophotometer

A good color-measuring reflectance spectrophotometer should have the following design features:

a. The instrument should be robust, shock- and vibration-proof and very simple to operate.

b. The measured colorimetric results obtained from the instrument should be bereproducible, i.e., the repeatability of the results should be within a tolerance limit. For this purpose, an instrument with a symmetrical double-beam optical arrangement with a 0/45 or 45/0 degree angle of incidence and reflectance is preferred.

c. The instrument should be able to measure colorimetric data with a desirable precision level of accuracy for both polychromatic and monochromatic measurements.

d. The sample holding attachment should have small areas of view and large areas of view in the instrument with different types of sample holders for fiber, yarn, fabric and other forms, as needed.

e. The instrument should have autocalibrated and autodiagnostic facilities with Calibrated white tiles and black tiles of known values of X, Y and Z.

f. The desired dataset in usable format for practical use should be easily obtainable from a computer without much modifications.

g. The instrument should be compatible with the latest version of computer operating systems.

h. The color measuring and match prediction software should be very user friendly.

i. The instrument should have the lowest cost and least maintenance requirement.

j. The instrument should not generate heat in the simple holding zone to avoid thermochromatic problems.

k. The transmittance mode should also be available optionally for optional measurements.

Color of a liquid or colored solution:

2.3.2.3 Important technical specifications of the reflectance spectrophotometer

The important technical specifications of a reflectance spectrophotometer are as follows.

a. Spectral Range:

The instrument's optics are designed so that they cover the spectral range from 400 to 700 mm for the measurement of color metric data at 10 mm intervals for general practice purposes. However, it is preferable to have a spectral range from 300 to 1100mm, with colorimetric data collected at 5 nm intervals.

b. Illumination and Viewing Geometry:

CIE has been recommended for geometries defining the direction of the incident light (incident angle of source light) and the direction of detecting the reflected light (the angle viewed by the detector placed appropriately), generally at 0°/45°, 45°/0°, near 0^0 or 8°/diffuse and diffuse/80°.

However, these four modes of viewing angles can be graphed into two major types: (i) bidirection types, i.e., having either 40/0 or 0/45 geometries (such geometries are most suitable for measuring samples with smooth surfaces, where the specular mirror reflection is usually exchanged), and (ii) integrated sphere types, where the sample is placed at one of the parts of an integrated sphere coated internally with Near 0^0 or 8°/diffuse and diffuse/0^0.

Geometries (In such geometry, the sample is illuminated either by diffuse light at all angles from the internal sphere, where the reflected light is viewed by the detector at 0° or near Zero 0° or above or near the normal to the sample surface (D/0 geometry); alternatively, the sample can be illuminated at or near Zero degree or at the normal to sample surface, and the reflected light is viewed diffusely (by 0/D geometry). D/o or near 0°, i.e., 8° or 0°/D geometries, can be used to measure samples with relatively nonsmooth sample surfaces, such as textiles or any textured samples. For smooth sample surfaces, an 8°/D integrated sphere-type geometry provides an optional measurement mode for including or excluding specular/mirror reflection by providing an adjustment facility for closing/opening a white specular port/component, with a black trap/hole along the 8° direction of the viewing/detection angle.

c. Light source, optics and optical alignment

Light Source: A light source/capable of producing light, equivalent to natural daylight, artificial daylight and fluorescent daylight is required for the study of extreme meteorism, etc. A continuous source of pulsed-xenon flash lamp is currently used for this purpose in color-measuring reflectance spectrophotometers.

Optics: Symmetrical double-beam optics are preferred as suitable for overcoming fluctuations in the intensity of the light source and the response error of photomultipliers by scanning the sample and comparing it with the standard simultaneously. Additionally, both polychrometic and monochromatic illumination/optical arrangements should be available, particularly for fluorescent samples, for which the detection of polychrometic reflectance is essential. For monochromatic measurements, appropriate dispersion elements, such as gratings, filters and continuous wedge interface filters, are used.

Optical Alignment: The autooptical alignment feature must be a modern color-remaining instrument.

d. Detector: Reflectance from the sample is measured in photodetectors using phototubes/photocells or photomultipliers, which also convert light output into electrical signals. An array of silicon photodetectors acts as an efficient defector and is preferred in modern color-measuring spectrophotometric instruments.

e. Other equipment modes: One of the other important equipment modes is the ability of small-area-of-view and large-area-view modes to measure color for small samples and larger samples, respectively.

Another mode may be the use of special sample holders (powder, fiber, yarn, etc.) or the provision of UV and l.R. cutoff filters.

The inclusion of the specular reflection component (SCI) and exclusion of the specular reflectance component (SCE) are other important modes for instrument use for specific purposes. An important feature of modern color measuring instruments is that they have options for autocalibration and autodiagnostic check facilities.

f. Special portable mode of the reflectance spectrophotometer: A specific

The development of on-line measurements of color in running production samples by a compact small spectrophotometer operated both by a battery and a line current, for more meaningful use of these spectrophotometers for on-site measurement of color in the production center for measuring color from distance, is possible for on-line control of color matching.

The microprocessor is based on a small screen and is a facilitator for obtaining minimum standard colorimetric data such as color strength, reflectance, witness, and yellowness indices. These data can be transferred to a computer that is suitably interfaced and has appropriate software support for further analysis and comparison for any other specific purpose.

Such a special reflectance spectrophotometer for on-line remote distance color measurement: Remote/distance measurement of color was not possible earlier with convention. Spectrophotometry, as an ordinary quartz and tungsten lamp light source, produces a light source of considerable/relatively greater duration, causing interference with ancient light, and the results are presented. However, the tenant of the pulsed xenon flash tube, which produces illumination of very short duration (20–30 μs) with very high radiant intensity (10 watts), has a higher signal/noise ratio, as does the use of modern diffraction grating with a detector array of parallel wavelengths (instead of a conventional holographic grating), which can produce a complete spectrum in parallel, where the use of a silicone photodiode detector array enables simultaneous measurement of all wavelengths parrlally and thar from a single pulse of illumination from moving objects or color data from distances from running colored textile fabrics. Online color measurements are now possible under normal light conditions without all other interference, such as air scattering.

Thus, with all the above discussed design features of color measuring spectrophotometers, computer color measuring and matching systems consist of the following:

a. A suitable reflectance spectrophotometer (preferably a double beam integrating sphere-type reflectance spectrophotometer).

b. A current generation computer with graphics design & printer integrated with the above-described reflectance spectrophotometer.

c. A suitable software version compatible with the current operating system for processing databases and color data for computer-aided color measuring and match prediction was developed.

Computer-aided color measurement and match prediction software is primarily based on the matchmetical conversion of basic reflectance data measured by a spectrophotometer to K/s values (as per Kubelka-Mark's equation), X, Y & Z tristimulus values or L, a, b values (as per C I E's formula) and the calculation of total color values on textiles by combinations of dyes (using the additive nature of K/s, value using Park & Steorn's algorithm Allen's metrix, etc.) for color matching recipe formulation, in addition to calculations of whiteness, yelloness and brightness indices as needed using relevant standard formulas.

Some of the eminent computer-aided color measuring and matching system (CCM System) manufacturing companies, which are widely used in India, are as follows:

i. An Applied Color System, USA.

ii. Macbeth, a division of Kollmorgan Co. USA & UK. (Presently Gretag-Mcbeth)

iii. Premier Color Scan Instrument Ltd., Navi -Mumbai, India,

iv. Instrument color system (ICS), UK.

v. Jay-Kay Pack CCM System by Jay Sinth Color Computer Co., India. Data Colur AG, Switzerland.

vi. Pretema, Switzerland.

vii. Hunter-Lab, USA.

viii. Milton Roy & Co. (Diano Color Products), USA and,

ix. Many others, including the BYK gardener in Germany.

2.3.2.4 Preparatory steps for executing color measurement and computer-aided color match prediction

The following are the sequential preparatory steps during its use for color measuring and color matching problem solutions with specific built-in software embedded in the textile industry and other industries:

2.3.2.4.1 Calibration of the instrument (reflectance spectrophotometry)

The calibration of the reflectance spectrophotometer before any new measurement is necessary, so the instrument provides accurate measurements of color values to predict and measure color with accuracy and a perfect scale of measurement.

Color is measured for dyed textiles, papers or leather goods, etc., at 400–700 nm in the visible zone and at 200–390nm in the UV zone (with the arrangement of UV on and off); hence, the instrument is calibrated against standard white tiles (giving a reading of 100% white) and then against standard black tiles (representing maximum absorption, i.e., 100% black). Therefore, with this white tile and black tile calibration, to accurately measure the color parameters on a perfect scale, the instrument scans a standard white tile and a standard black tile, and a reflectance spectrophotometer is used to adjust the reflectance values to the proper scale to obtain accurate measurements.

2.3.2.4.2 Quality control checks for dyed textiles and apparels

In the quality control option, the color data of any sample can be measured and stored for comparison of different color parameters between standard samples and a newer incoming batch produced and thus can be compared against color parameters of any set of existing standards to measure color difference values, variations in color tone with color difference space diagrams, etc., in addition to reflectance and K/S values and metamerism indices, etc., to determine whether those color values are within a tolerance limit, i.e., whether the sample is passed or failed in quality checks for acceptance and rejection decisions.

Thus, a colorimetric analysis of any two dyed/colored samples (produced and standard sample) can be performed in terms of their difference in predominant hue and their L, a, b color differences in terms of lightness/darkness, redness/greenness, yellowness/blueness to understand and correct the tonal variation with determination of color differences by saturation/chroma value and finally by evaluation of the overall position of the L, a and b color coordinates or position of the coordinates of the X, Y and Z tristimulus values in the color space diagram.

The quality control option in the computer-aided color measuring and matching software menu provides graphical plots showing the representation of dye concentration vs. reflectance, dye concentration vs. K/S values, and all graphical representations of any color parameters against 400–700 nm wavelengths in the visible region to determine the predominant wavelengths where the absorption maxima of the color parameters/values are obtained for the measured samples in Ref./compared with any standard sample.

There are options for determining the whiteness index, yellowness index, brightness index, and metamerism index between any two colored samples (produced and standard) when these data are important for quality checks of any colored sample.

2.3.2.4.3 Color measurement parameters

The following are the important color parameters, which can be evaluated using a calibrated reflectance spectrophotometer attached to a CPU processor and dedicated software to process and store the color data useful for color matching of any textile product:

i. *The reflectance values (R)/X, Y and Z are the Tristimulus values of any colored* textile. The basic measured values from the UV–VIS reflectance spectrophotometer are either the Reflectance values or the Tristimulus values of the colored substrate, which can later be converted to any other color parameter using proper/relevant colorimetric relationships with the respective CIE lab-1976 relevant equations/formulae [2, 3]

ii. *Color Differences:* Color differences between any two samples (standard and produced) were evaluated by analysis of the respective color data observed and graphical plots of color difference analysis.

Color Strength: To evaluate the surface color strength, i.e., the K/S values of a colored sample.

Indices: To evaluate different color indices, such as the whiteness, yellowness, brightness indices and metamerism indices.

Combined Output: To evaluate all color parameters by measuring the reflectance of colored samples in terms of surface color strength and color difference, all graphical plots were combined for analysis of the output results at a glance on a single screen.

Color database: A standard color database is prepared and stored for each dye class for each separate textile substrate for use in predicting color matching recipes.

2.4 Major application areas for computer-aided color measuring and matching systems

The following are the major application areas for computer-aided color measuring and matching systems.

1. The following color parameters were measured: the reflectance (R) and the resistivity (X, Y and Z) at λ_{max} (at the maximum absorbance wavelength) were measured, the K/S values were calculated, and all the color data were stored for further analysis.

2. The color differences were calculated by the CIELab*-1976 formula for determining the total color difference values, i.e., $DE^* = \sqrt{[DL^*)^2 + (Da^*)^2 + (Db^*)^2]}$, for comparison of the color differences in terms of the L*, a* and b* scales and plotting the data in a color space diagram in terms of the L, a and b coordinates, i.e., the DL*, Da* and Db* values in addition to the DC^*_h, MI, etc. A quality check of the newer shades of the produced dyed textiles for shade sorting revealed the deviations of the three-coordinate color difference from the central point in terms of the lightness/darkness (DL*), redness/greenness (Da*) and blueness/dieslowness (Db*) color variations with the least possible metameric values (MI-Labd scale).

3. The color database for particular classes of dyes (classwise and companywise) applicable for particular substrates was prepared and stored, and the degree of linearity/nonlinearity was checked by checking plots of dye concentrations vs. K/S values and linearization of those color databases before storing those data as standardized color database files by curve fitting with least square or moving average methods for use in the prediction of computerized color matching formulations to obtain the least expensive and least metameric color matching formulations using previously stored specific types of color databases of different classes of dyes for different companies for specific substrates.

4. To carry out color matching, recipe correction by the manual batch correction method or by the computerized autocrection method is used to obtain precision matching of shades with the least metameric deviations.

5. The calculations were related to the use of dye waste liquor with the remaining color in the waste solution by further decreasing the quantity of dye added to the waste liquor.

6. Prediction of % purity and checking of the quality of the incoming new batch of dyes with the help of transmittance mode.

7. Calculations of indices such as the whiteness index, yellowness index, and brightness index of scoured and bleached textiles at corresponding standard scales following the relevant CIE, Hunter Lab, ASTM-E-313, Stansby/Burger and ISO-2470/Tappi scales were performed.

8. To determine the efficacy of action of optical brighteners or optical bluing agents (OBA) utilizing a computer-aided color measuring and matching system, a particular color dataset was measured with and without UV light in on and off mode, and the results of the peak value in the UV zone and visible zone in UV-in mode and UV-off mode were compared.

9. The quantitative prediction of color fading behavior after washing/crocking/ perspiration and exposure to UV light, rather than the conventional visual assessment of loss of depth and staining scale, was performed according to the precision rating of color fastness grades instead of the use of the grayscale rating option.

10. To determine the efficacy of a detergent/surfactant for soil removal, reflectance values were measured before and after washing the soiled clothes under preset washing conditions.

In addition to the above applications, several advanced applications of the UV VIS absorbance colorimetric/photometric spectrophotometer (by using both an absorbance spectrophotometer and a reflectance spectrophotometer) include the following:

i. Identification of specific color constituents of any natural or synthetic color/ dye/pigment by UV–VIS spectral scan and its peaks for wavelength vs.

absorbance as fingerprints for particular color constituents in a dye or pigment.

ii. To study dye compatibility by analyzing the dyeing rate or progressive color buildup under various dyeing conditions—either by (a) varying the dyeing temperature and time profile or by (b) varying dye concentrations while keeping all other dyeing parameters fixed—and then comparing plots of DC vs. DL and by plotting DL vs. K/S vs. DL and by (iii) applying varying proportions of two dyes for compatibility tests and analyzing their color difference index values to determine differences in the CDI for different proportions of dyes used.

iii. To study the effects of variations in dyeing conditions (such as dye concentration, mordant concentration for natural dyeing, dye bath pH, dyeing time, dyeing temperature, MLR, and salt concentrations)/any other dyeing additive, the dyeing variables/dyeing process parameters were standardized to obtain the optimum color yield.

2.5 Some practical considerations for accurate measurement of color

The practical aspects of preparing accurate color databases and their linearization for subsequent use in generating newer color matching formulations/recipes using stored standardized color databases are too important. Equally important is setting up proper tolerance values for DE and multiple color difference tolerances in terms of DE*, DL*, Da*, and Db* values under a specified known standard illumination (e.g., D65 or artificial day light, fluorescent light). Accurate measurement of the color data set in a UV–VIS reflectance spectrophotometer depends on the following factors; hence, care during measurement of the color data set should be taken to avoid accurate measurements:

a. Background Opaqueness of the samples mounted by multiple folding of the colored samples.

b. Uniformity in actual dyeing (not exceeding 5% CV of K/S values).

c. Sample orientation (unidirectional, either warp or weftwise prefixed).

d. Changes in the surface structure or scattering of the surface of the sample are avoided.

e. The two-sided nature of the sample may have affected the results.

f. Slabbing or dyeing defects should be avoided.

g. Separate measurements were required for fluorescent colorants.

h. Separate factors to include for dull shade, bright shades, etc.

i. Accurate shades may arise due to inaccurate weights of dyes and chemicals.

3. Some case studies for quality checks of colored textiles and dyed apparel products

3.1 Identification of color constituents (based on the ingredient index) from dyed textiles

Natural dyes such as Indigo can be identified and distinguished from synthetic indigo from natural indigo-dyed cotton textile fabrics by extracting the color component, followed by UV–VIS spectrophotometry. The test method [22] standardized by the BIS to distinguish synthetic indigo from natural indigo is as follows:

First, 0.1 g of each of the two types of indigo dyes (natural indigo and synthetic indigo) was dissolved in 1000 ml of dichloromethane, diluted 5–10 times and scanned through a UV–visible spectrophotometer to obtain the WL vs. absorption plots. In the visible zone, identification of the natural indigo dye can be confirmed from the values of the lambda maxima (absorption/optical density) of both natural indigo and synthetic indigo and by comparison of the characteristic peaks in both plots, as shown in **Figure 2**. **Table 1** summarizes the observed peaks in tabular form in the UV–visible spectra of natural indigo and synthetic indigo.

Thus, the abovementioned UV–vis spectroscopy-based colorimetric analysis helped with the corresponding UV–vis plots to help identify natural indigo compared to synthetic indigo from colored textiles.

3.2 Verifying the linearity of the plots of dye concentrations vs. K/S values for the stored color database

Kubelka and Munk's functions [2–5] do not always exhibit a satisfactory linear relation between dye concentration and K/S plots; therefore, a number of empirical formulas, such as color difference index (CDI) [15–17] values or modified magnetic relationships according to K–M theory [2, 3], have been proposed and subsequently used with caution to achieve accurate shade percentages during the preparation of color databases for storage.

One of the deficiencies of K–M theory is that the optical discontinuity for the intervention of air vs. sample interfaces is not taken into consideration when

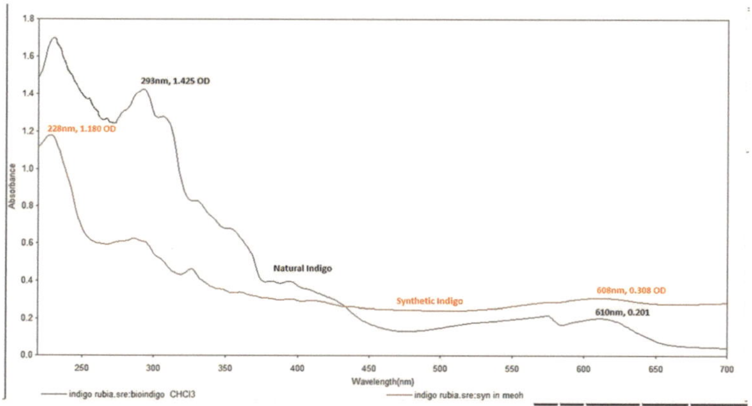

Figure 2.
UV–vis spectra of natural indigo and synthetic indigo.

Peak value (nm)	For Synthetic Indigo	For Natural Indigo	Remarks on observed differences In peak value and type
220–295	228 nm (1.18 OD)	293 nm (1.425 OD)	The sharp peak at 228 nm is clear characteristic of UV–VIS peaks of synthetic indigo
605–611	608 nm (0.20 OD)	610 nm (0.201 OD)	The lambda max peak for natural indigo is shifted to 610–611 nm from 608 nm (for synthetic indigo) due to natural contaminants and association of other natural minor matters/constituents in natural indigo.

Table 1.
UV–VIS spectral plots of natural and synthetic indigo.

formulating K–M theory. This optical discontinuity results in the reflection of light from the surface of the specimen kept open in air without even interacting with dye molecules.

A number of theoretical techniques have been proposed to minimize the error introduced in recipe calculations due to the nonlinear relation between the K–M function and the concentration of dyes. If a sufficiently high exhaustion of dye can be achieved at a high level of dyeing, approximately good linearity may be obtained. The simplest way to determine the optical data is to take the average of all absorption coefficient values obtained for specimens at six to eight or eight to ten levels of dyeing during calibration dyeing to prepare/generate a color database for a particular dye class for a specific textile substrate.

For the preparation of a standard color database (Dye Casswise) for computer-aided color matching databases, the following factors are considered for least metameric color match formulations:

i. No layers of folded dyed fabric were used to maintain the uniformity of the produced dyed sample for measuring reflectance and K/S color data for all samples of dyed fabrics.

ii. The orientations of the sample are either warp or weft wayall throughout the measurements for the produced dyed samples.

iii. Predecided standards for level/unlevel decisions in dyeing interms of the dyeing uniformity value expressed by the % coefficient of variation (CV) for K/S values (within 5% CV for K/S values at different points of colored fabric is considered to indicate uniform dyeing).

iv. Color matching specifications of tolerance values for multiple color difference scales (in terms of DE*, DL*. Da* and Db* are predetermined for comparing and predicting color matching formulations. Instrumental data vs. human eye-perceived color differences for match/nonmatch decisions with predecided tolerances, i.e., predecided tolerances of DE* (some where also shown/designated ΔE, both are the same), should be within 0.5 to 1.0, with optional limits of DL*, Da*, and Db* tolerances. DE* values above 1.0 are not generally considered acceptable matched samples.

v. The linear/nonlinear behavior of the plot of dye concentration vs. K/S can be expressed as

After the produced sample was dyed for color database preparation, the sample was checked. Therefore, for all dyed samples, plots (calibration curves of the color database for each dye) of dye concentration vs. K/S must be set to be linear by appropriate curve fitting/linearization processes. If such a calibration curve for any dye is not linear, it can be linearized by the least square linear curve fitting technique or by eliminating a few points/concentrations of dye to make it linear to be stored as a color database for color match prediction for particular fabric and dye class combinations.

With the above point (v) in mind, all color database results of the produced samples are shown in **Table 2** and were checked for linearity of dye concentration vs. K/S plots before entering and storing these color databases for each dye for predicting precision color match formulations with that particular selective class of natural/synthetic dyes.

For practical examples of such linearization of the color database, the following example of a simple case study for linearization of the color database produced, for example, for one selective dye, is shown in **Table 2,** along with the plot of dye concentration vs. K/S values in **Figure 3**, where linearization was performed, as

Name and ID of the Dye	Dye conc,	K/S (%)	CV% of K/S Value	DE	DL	Da	Db	DC	ΔH	MI	CDI
Dye-1	0.5	5.07	3.86	6.15	4.44	0.13	4.25	4.63	0.13	2.62	0.07
	1.0	6.03	4.24	7.05	4.90	0.16	5.06	4.08	0.15	2.84	0.07
	1.5	6.49	4.14	5.44	3.81	0.46	3.86	4.04	0.42	2.48	0.24
	2.0	6.75	3.90	3.83	3.85	1.21	2.99	3.72	1.08	2.2	0.62
	2.5	7.96	356	6.23	4.19	0.62	4.57	4.32	0.54	2.57	0.28
	3.0	9.75	3.67	6.17	3.59	0.96	4.92	4.00	0.82	2.63	0.39
	3.5	10.49	4.02	5.20	2.84	1.78	2.45	4.15	1.61	1.95	1.04
	4.0	10.54	5.09	5.10	3.00	0.66	4.07	4.57	0.38	2.26	0.21
	5.0	10.87	5.13	5.07	2.98	1.91	4.26	4.51	1.63	2.34	0.81
Dye-2	0.5	4.67	3.60	5.81	3.90	1.84	3.88	12.63	1.60	2.67	0.87
	1.0	4.71	3.41	5.72	4.14	0.97	3.83	14.33	0.87	2.46	0.53
	1.5	5.33	3.40	5.82	3.86	1.89	3.92	14.64	1.65	2.69	0.88
	2.0	5.34	3.83	5.27	3.15	2.30	3.55	15.35	2.02	2.7	1.06
	2.5	5.80	3.90	5.03	3.12	3.28	3.17	15.03	2.89	2.82	1.46
	3.0	5.90	3.10	6.23	3.87	2.18	4.37	15.46	1.86	2.92	0.88
	3.5	6.01	3.37	5.05	3.50	3.07	3.14	15.57	2.71	2.81	1.41
	4.0	6.32	4.18	5.55	2.52	3.40	3.59	15.73	2.96	3.05	1.35
	5.0	6.93	4.12	5.64	2.53	3.60	3.53	15.51	3.77	3.13	1.43

Table 2.
Color database results for two selected dyes applied to the produced samples for calibration dyeing for preparing a color database for precision color matching.

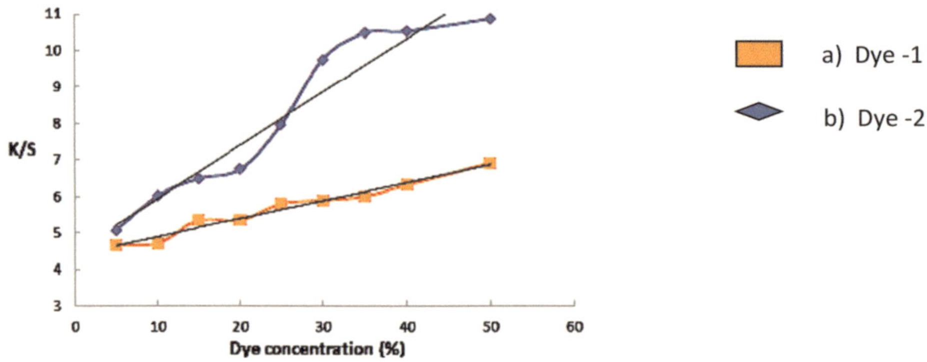

Figure 3.
(a) And (b) linearization of K/S vs. dye concentration (0.5%–5.0%) for achieving accuracy in color match prediction in a database for 2 different dyes.

shown by the other broken lines following curve fitting performed by the least squares method in **Figure 3.**

Figure 3 shows that Dye 1 (orange line) had an almost linear relationship with the Dye concentration vs. the K/S plot, with minimum variation and lower ranges of the CV % of the K/S value and greater ranges of DC values (**Table 2**) compared to those of Dye-2, showing much greater linearity (blue–violet line) for the Dye concentration vs K/S plot (**Figure 3**). Hence, the linear curve fitting principle must be applied in the case of Dye-1, as shown by the extra fitted linear curve shown therein, and the linearized data must be stored as a color database. Similarly, more dye calibration data for preparing a total color database for particular classes of dyes will be needed.

3.3 Study of the compatibility of dyes for producing compound shades

There are many methods to test compatibility between any two dyes, and hence, compatibility can be determined in different ways. However, only two methods are popular: (i) the conventional method of comparing the color build-up rate by dyeing under two different SETs—by varying the dyeing time and temperature profile in one set and by varying the dye concentrations in another set—and then to compare plots of K/S vs. DL and DC vs. DL, where two dyes under the test reference showing the degree of similarity of the rates of color build-up in these two sets are more compatible.

Recently, another easier and simple colorimetric method called the relative compatibility rating method [15], which is based on calculations of differences in the CDI of dyes with various proportions of any two dyes and can be used to rate compatibility on a 1–5 scale, was established. Both of these methods are shown as examples for dyeing cotton with direct dyes, as given below:

3.3.1 Conventional methods for dye compatibility testing

This method is based on a comparison of the gradual color build-up rate for dyeing under two sets of variations during dyeing for gradual color build-up, i.e., (i) SET-1, by varying the dyeing time and temperature profile (keeping the dye concentration and dyeing process variables fixed), and (ii) SET-II, by varying the overall dye concentration of the 50;50 binary mixture of any two dyes (keeping the dyeing time and temperature and other dyeing parameters fixed). The data for 4 RBGY colored selected binary pairs of direct dyes taken for dyeing a preselected cotton textile substrate are shown here in **Tables 3–5**.

The DL vs. K/S and DL vs. DC plots of the corresponding data from the above **Tables 3–5** for Set-I and Set-II indicate no particular trend of the color build-up rate for M1 (direct red color 1 and direct green color-2) for both sets, and these two dyes are therefore poorly compatible. For M2: Direct Green Color-2 and Direct T. Blue Color-3.

Similarly, both Set-I and Set-II had moderately compatible colors, and M3 (direct green color-2 and direct yellow color-4) had increasing colors in the same direction in both Set-I and Set-II; however, these colors varied here, and they were judged to be fairly compatible, as per the conventional test of dye compatibility, as understood

Dyeing profile	SET-I: At Varying Dyeing Time/ Temperature profile (min, °C)					SET-II: At Varying dye concentration (in parts) and prefixed other conditions of dyeing					
Dye-1: Dye-2 50:50	**K/S**	**ΔL**	**Δa**	**Δb**	**ΔC**	**Dye parts**	**K/S**	**ΔL**	**Δa**	**Δb**	**ΔC**
Undyed	0.53					0					
50 °C, 10 min	1.82	15.7	10.6	13.8	7.83	10	1.33	10.9	0.57	11.1	4.52
60°C, 20 min	1.88	14.4	7.41	13.4	2.34	20	1.73	12.1	0.75	12.8	2.75
70°C, 30 min	3.81	15.4	8.72	13.9	3.15	40	2.27	12.5	0.36	12.9	2.71
80°C, 40 min	3.76	12.0	1.2	12.0	3.51	60	3.13	13.4	0.109	13.2	2.36
90°C, 50 min	3.87	12.5	2.8	11.8	3.10	80	3.82	13.5	1.082	12.9	2.47
100°C, 60min	3.62	11.6	3.0	11.1	3.59	100	5.53	13.9	3.37	12.8	1.56

Table 3.
Color parameters under two sets of color build-up experiments for testing dye compatibility for selected binary mixtures of two dyes (50:50): M1: Direct red color-1 and direct green color-2.

Dyeing profile	SET-I: At varying dyeing times/SET-II: At varying dye concentrations (in parts) and temperature profile (min, °C) prefixed for other conditions of dyeing										
Dye-2: Dye-3 50:50	**K/S**	**ΔL**	**Δa**	**Δb**	**ΔC**	**Dye parts**	**K/S**	**ΔL**	**Δa**	**Δb**	**ΔC**
Undyed	0.53					0	0.53				
50 °C, 10 min	1.4	8.63	13.5	9.65	5.49	10	1.22	8.55	14.02	9.51	5.91
60°C, 20 min	1.75	8.58	13.7	9.30	5.57	20	1.75	9.78	14.36	10.3	6.37
70°C, 30 min	2.06	9.58	14.8	10.4	6.84	40	1.98	9.96	15.82	10.9	7.92
80°C, 40 min	2.62	9.79	15.0	11.1	7.19	60	2.53	9.84	15.32	10.6	7.36
90°C, 50 min	2.72	9.97	15.2	10.7	7.34	80	3.48	9.89	14.79	10.3	6.78
100°C, 60min	3.38	10.1	15.2	10.5	7.19	100	3.63	10.3	14.93	10.9	7.05

Table 4.
Color parameters under two sets of color build-up experiments for test of dye compatibility for selected binary mixtures of two dyes (50:50) M2: Direct green Color-2 and direct T. Blue Color-3.

Dyeing profile	SET-I: At varying dyeing times/SET-II: At varying dye concentrations (in parts) and temperature profile (min, °C) prefixed for other conditions of dyeing										
Dye-2: Dye-4 50:50	**K/S**	**ΔL**	**Δa**	**Δb**	**ΔC**	**Dye parts**	**K/S**	**ΔL**	**Δa**	**Δb**	**ΔC**
Undyed	0.53					0	0.53				
50 °C, 10 min	5.61	1.29	9.97	9.14	11.6	10	2.68	1.05	10.27	3.85	7.47
60 °C, 20 min	5.96	1.35	10.0	7.45	10.2	20	3.62	1.35	10.96	3.86	7.93
70°C, 30 min	6.36	1.97	11.2	4.74	8.77	40	4.52	2.08	11.46	3.43	7.95
80°C, 40 min	8.37	3.76	10.9	0.99	5.96	60	5.56	3.45	11.24	1.06	6.19
90°C, 50 min	8.53	4.06	11.2	0.39	5.33	80	6.51	3.13	11.29	1.38	6.43
100°C, 60 min	4.47	6.25	10.9	2.87	3.86	100	8.93	3.80	10.64	0.04	5.08

Table 5.
Color parameters under two sets of color build-up experiments for test of dye compatibility for selected binary mixtures of two dyes (50:50) M3: Direct green color-2 and direct yellow color-4.

from **Figure 4.** Compatibility tests of K/S vs. DL and plots of DC vs. DL for each binary pair of dyes, M1: Direct Red color-1 and Direct Green color-2, M2: Direct Green color-2 and Direct T. Blue color-3 and M3: Direct Green color-2 and Direct Yellow-4 pairs of direct dyes applied to cotton, are shown below:

3.3.2 Newer RCR method for dye compatibility testing

In this newer and simple method [15–17], the dyeing of selected fabrics was carried out with each mixture of binary pairs of selected dyes in varying proportions, and the colorimetric measurements are shown in **Table 6.** The differences in the CDI values for different proportions of dyes used are shown in **Table** 7, where the relative compatibility.

Ratings/grades for each binary pair of selected dyes are assigned/graded according to the chart value of the rating (as shown in **Table 8** as per earlier established chart values [15–17]) with respect to the maximum differences in the CDI values obtained, as shown in **Table** 7. Both conventional methods and simple newer RCR methods for compatibility testing between selected dyes showed the same or similar results, proving the equal efficacy of the newer RCR method for dye compatibility tests; hence, this method may be considered useful for industry because it is quicker and easier to use than other conventional methods.

3.4 Computer-aided color match prediction data and its practical use

For the prediction of computer-aided color matching formulations for any or different known and unknown standard samples, different color matching formulations can be easily obtained after setting allowed/permissible color matching alternatives under specific illuminants and under specific instrumental setups, resulting in different levels of metamerism and cost.

Among all the predicted color matching formulations generated, the least cost and least metameric match results are identified for practical use. One such match

DOI: http://dx.doi.org/10.5772/intechopen.114181

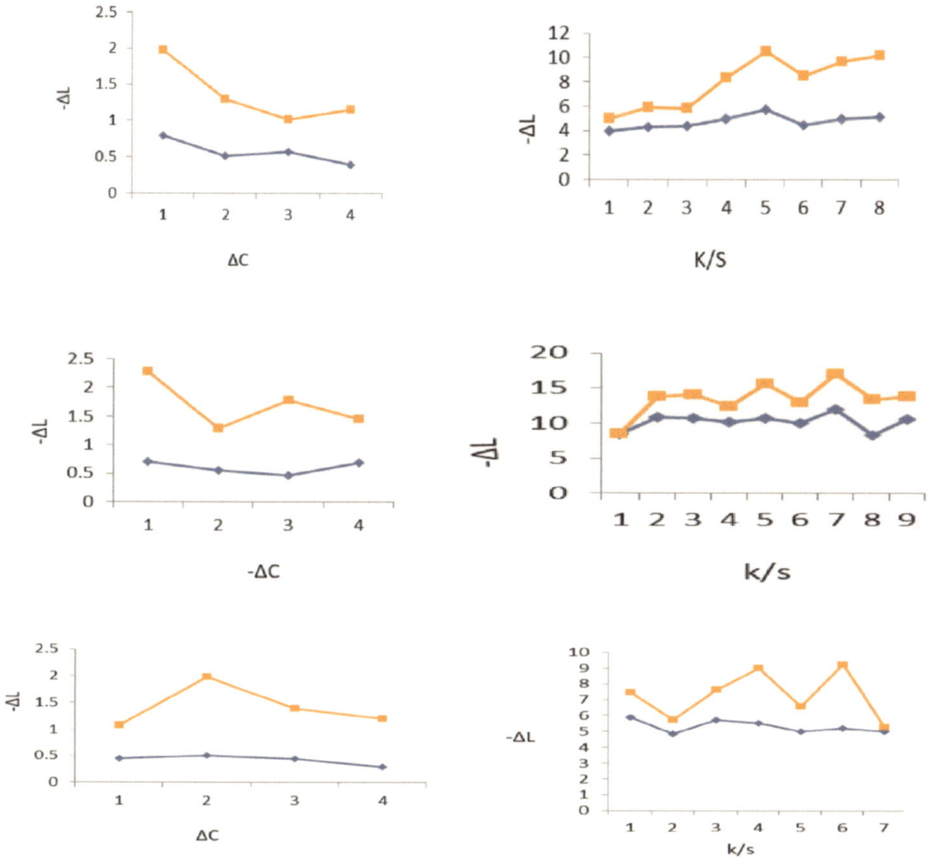

Figure 4.
DC vs DL and K/S vs DL plots for dyeing results under two dyeing conditions, set-I and set-II, for determining the progressive color build-up rate to test the compatibility of dyes in the conventional colorimetric analysis method for M1: Direct red color-1 and direct green color-2; M2: Direct green color-2 and direct T; and blue color-3 and M3: Direct green color-2 and direct yellow color-4.

prediction formulation is shown in **Table 9**, where formulation 1 is the least metameric and formulation 2 is the least expensive.

After one formulation (e.g., Formulation 2 as the least cost recipe) from among the two or more formulations generated, actual/practical lab dyeing must be performed with the predicted percentage of the mixture of dyes/colorants to check the practical match results of the produced samples by comparing their color parameters, for a particular match of the standard sample (C-25), after practical dyeing in the laboratory and then in the bulk, and then to compare the predicted and actual practically measured color difference parameters of the produced samples against the standard sample given to match.

Table 10 contains comparative color data for further analysis of predicted formulations vs. actual dyeing results for checking the acceptability of matching formulations with newer color matching indices, as shown below.

The higher the color matching index (CMI) is, the greater the deviation of the color parameters of the predicted computer-aided color match results from the results of the actual dyed sample, i.e., the lower the efficiency of the computer-aided color match-predicted formulations. Among the two predicted formulations generated in

Dyes/combination	K/S at λ_{max}	CV % of K/S	ΔE	ΔC	ΔH	MI	BI	CDI
For 75:25 ratio of each color/dye in each of M1,M2 and M3 binary pairs of dyes taken								
M1 (D Red color-1:D Green color-2)	5.76	5.10	21.02	−14.44	15.12	4.41	9.88	4.99
M2(D. Green color-2:DT.Blue color-3)	7.66	3.67	20.03	14.49	16.74	5.07	9.43	4.56
M3(D. Green color-1: D Yellow color-4)	10.08	5.13	21.08	13.27	10.34	4.02	6.92	4.08
For 50:50 ratio of each color/dye in each of M1,M2 and M3 binary pairs of dyes taken								
M1(D Red color-1:D Green color-2)	6.71	8.37	20.16	−14.87	13.87	4.31	6.69	4.36
M2(D. Green color-2:DT.Blue color-3)	8.00	6.25	21.99	16.53	18.32	4.89	9.93	4.80
M3(D. Green color-1: D Yellow color-4)	10.54	7.13	18.53	14.10	18.47	5.23	5.43	4.64
For 25:75 ratio of each color/dye in each of M1,M2 and M3 binary pairs of dyes taken								
M1(D Red color-1:D Green color-2)	7.19	2.13	19.47	−14.33	12.82	5.05	8.37	3.44
M2(D. Green color-2:DT.Blue color-3)	8.68	3.67	22.30	16.86	18.56	5.15	10.89	4.76
M3(D. Green color-1: D Yellow color-4)	11.58	8.39	21.14	14.61	15.59	5.69	4.86	3.94

Table 6.
Colorimetric analysis of data for testing dye compatibility by the RCR method by dyeing any two selected dyes in different proportions for cotton fabrics dyed with each binary mixture of direct dyes in three different proportions.

Binary pair of Dyes combination	CDI values			$CDI_{max} - CDI_{min}$ difference	RCR From chart	Compatibility grade as per RCR values from chart
Prportion of dyes	75:25	50:50	25:75	—		
M1(D Red color-1: D Green color-2)	4.99	4.36	3.44	4.99–3.44 = 1.55	2	Poor
M2(D. Green color-2:DT.Blue color-3)	4.80	4.98	4.76	4.98–4.76 = 0.22	3–4	Moderate
M3(D. Green color-1: D Yellow color-4)	4.08	4.64	3.94	4.64–3.94 = 0.70	2–3	Fair

Table 7.
*The color difference index (CDI) and maximum differences in the CDI values [$CDI_{max} - CDI_{min}$] were compared with the charted relative compatibility rating/grade (as per the RCR values charted in **Table 8**) to determine the compatibility rating.*

Table 9, two predicted formulations vs. actual dyeing with formulations 1 and 2 are compared:

Table 10 indicates that a closer match is obtained by Formulation 1, which is the least metameric, with a CMI of 3.02 compared to the CMI of 4.05 for formulation 2,

Compatibility Grade	RCR values	Differences between maximum CDI and minimum CDI values among all dyeing results for use of all proportions of binary pair of dyes taken for dyeing *
Excellent	5	>0 but ≤0.05
Very Good	4–5	>0.05 but ≤0.10
Good	4	>0.10 but ≤0.20
Moderate	3–4	>0.20 but ≤0.30
Average	3	>0.30 but ≤0.40
Fair	2–3	>0.50 but ≤1.00
Poor	2	>1.00 but ≤5.00
Very Poor	1–2	>5.00 but ≤10.00
Worst	1	>10.00 but ≤15.00
Non-compatible	0	> > 15.00

**Source: [15].*

Table 8.
Chart values [15] for relative compatibility rating (RCR) vs compatibility grade.*

which is the least expensive formulation. However, as CMI values for both formulations are within 5.0, they are considered acceptable matches. However, for any predicted formulation, after practical dyeing, with a CMI greater than 5, there is a reasonable difference/deviation in color parameters after actual dyeing against the predicted formulation; hence, this difference should be corrected by batch correction tools until a more precise match with a CMI within 5 is achieved.

3.5 Effects of variations in the concentrations of minerals (for natural dyeing) and dye/pigment colored components and other dyeing process variables

To understand the effects of mordant concentrations and dye concentrations and other dyeing conditions/dyeing process variables may be studied by varying one.

The other parameters of the dyeing variables are fixed – by varying one by one (e.g., the dye concentrations are varied, while the other dyeing variables are fixed and held constant), and the observed dyeing results can be assessed/compared to determine which value of the input parameter of the dyeing process (Say– dye concentration) yields the optimum color yield or maximum K/S value, and that value of the dyeing parameter is taken as the optimum to standardize it, as per the data shown in **Tables 11** and **12,** for the variation of all the different parameters of a specific dyeing process for a particular combination of mordant and dye for a specific natural dyeing process. to standardize the staining conditions. For example, in a specific case study, for overall varying concentrations of Harda (H) and Potash alum (Al), which are combined dual mordants applied in sequence on cotton, and for varying dye concentrations (2–8% purified catechu extract dye), the observed dyeing results for varying dyeing parameters are shown in **Table 11**.

The effects of varying other dyeing process parameters and other dyeing conditions, such as dye bath pH, time of dyeing, temperature of dyeing, MLR of dye bath,

Standard Name		=	C-25 Cotton	Formulation report				Mode = R	Spectro-RFL	
R	2.28	2.50		3.1287	4.09	4.74	4.27	4.41	4.71	5.08
6.19	5.48	5.08	4.82	4.512	4.31	4.07	3.88	3.79	3.61	3.33
File name: AKS-2			Database: AKS-Cotton-Direct Dye Set-3				Nos. of Dyes: 3 dyes combination			
Dye ID# selected		All	=	D-1, D-2, D-3, D-4, D-5, D-6, D-7, D-8, D-9, and D-10					On cotton	
DE*, D65'. 1.00		Tolerence of DE-value =			1.00 t = 100		Batch Amount = 100 kg			
Dye Exhaustion factors (included)			1.0	Substrate ID# 3, Substrate Name. Bleached cotton- 3 & Substrate factor = 1.0						
Colorant ID# and Recipe/ Formulation		Dye Cost	Amount for 100 kg batch and also in %.	da*	db*	dL* ClE		dE*	Rs (Dye Cost)	
Formula#1		(Rs/kg)				Least metameric match				
1	Dye-1	320	10.0 (1.0%)	D	−0.11	−0.10	0.04	0.05	32.00	
5	Dye-5	400	18.5 (1.85%)	A	−0.61	0.20	0.0	0.21	74.00	
9	Dye-9	390	07.0 (0.07%)	F	−0.42	0.31	0.11	0.52	27.30	
	Total dye		35.5 (2.92%)						133.30	
Formula # 2						Least cost match				
1	Dye-1	320	8.0 (0.08%)	D	−0.40	−0.30	0.0	0.51′	25.60	
2	Dye-2	300	12.5 (1,25%)	A	−0.52	0.31	0.0	0.61	37.50	
6	Dye-6	350	13.5 (1.35%)	F	−0.50	0.31	0.32	0.72	47.25	
	total dye		34.0 (1.68%)						110.35	

***Note = This color matching formulation/recipe was generated in a computer-aided reflectance spectrophotometer (Make: Premier Color Scan Pvt. Instrument Ltd., Model SC 5100A, associated with Color-Lab plus software) using a formulation menu and direct dye database prepared in the laboratory for color match prediction for cotton with a predetermined tolerance of ΔE^* within 1.0.*

Table 9.
Predicted color matching formulations were generated using a specific color database stored for a specific substrate (cotton) for a specific dye class from a specific company.

and salt % added to the dye bath, on the color yield/surface color strength are shown in **Table 12.**

All other dyeng process variables were held constant, with the variation of the abovementioned dyeing conditions one after another for natural catechu dyeing on cotton, with 10% overall application of alum plus hardness (50:50) applied in sequence, followed by 6%.

The aqueous extract of the purified catechu dye powder gave the best/maximum K/S values and was more uniform. With the mordant and dye concentrations fixed, the effects of other dyeing conditions/dyeing process variables, such as dye bath pH, time of dyeing, temperature of dyeing, MLR of dye bath, and salt percentage, on the color yield/surface strength show maximum/optimum color values, which are considered the optimum dyeing conditions (**Table 12**) for the use of a dyeing time of 60min and a dyeing temperature. At 70°C and a dye bath pH of −4-5 (acidic) with an

DOI: http://dx.doi.org/10.5772/intechopen.114181

Sample ID	Dye used/and type of fabric	K/S at 420 nm	CV % of K/S	ΔE	ΔL	Δa	Δb	ΔC	ΔH	MI	CMI
Sample C25-Std Sample	(Standard Dyed cotton)	6.98	5.07	55.17	−42.88	16.84	30.27	8.92	−22.98	5.62	—
Predicted Formulation −2 (vide Table-5)	No actual Dyeing, predicted data-	—	—	−.51	−0.41	0.30	0.0	—	—	—	—
Predicted Formulation −1 (vide Table-5)	No actual dyeing, predicted data	—	—	0.50	0.04	−0.11	0.10	—	—	—	—
Dyed-with Formulation-2 (Least cost)	Actual dyed with F-2	6.80	4.17	0.90	0.35 (darker)	−0.25 (greener)	−0.21 (bluer)	7.30	−20.47	7.61	4.05
Dyed-with Formulation-1 (least-metameric)	Actual Dyed with F-1	6.68	4.57	0.50	0.40 (Less darker)	−0.30 (less greener)	−0.0 (Still bluer)	7.25	−19.47	6.91	3.02

Table 10.
Comparative color data of formulation 2 and formulation 1 by determination of color matching index (CMI) after actual dyeing for matching against standard given colored samples for checking match acceptability.

Sample	K/S	ΔE	ΔL	Δa	Δb	ΔC	ΔH	BI	MI	CDI
Control	0.055							80.12		
H-10%- Cat-2%	3.28	50.45	−36.96	17.09	29.77	25.66	−22.81	9.63	1556.37	0.028
H-10%-Cat-4%	3.98	49.45	−36.06	16.19	28.97	24.76	−22.20	9.33	1520.37	0.030
H-10%-Cat-6%	4.77	48.38	−36.11	15.61	28.16	23.48	−22.03	9.15	1516.19	0.029
Al −10%-Cat-2%	4.57	49.97	−36.83	17.87	28.67	25.25	−22.44	6.74	1548.97	0.027
Al −10%- Cat-4%	4.81	50.21	−38.49	15.71	28.15	23.53	−22.03	6.02	1588.25	0.029
Al −10% - Cat-6%	4.98	56.12	−40.88	15.84	29.27	23.90	−22.08	5.98	1665.94	0.028
Al-H-5%, + Cat-2%	4.46	48.95	−38.58	15.62	25.76	21.57	21.02	10.23	1583.57	0.030
Al-H-10% +,Cat-2%	4.57	49.97	−36.83	17.87	28.67	25.25	−22.44	10.24	1548.97	0.027
Al-H-15% +,Cat-2%	5.19	50.94	−39.49	16.66	27.51	23.59	−21.36	10.27	1604.94	0.028
Al-H-20% +,Cat-2%	5.46	51.73	−41.53	16.51	26.05	22.36	−21.24	9.16	1641.49	0.029
Al-H-5%, + Cat-4%	5.78	52.73	−41.37	17.00	27.92	24.14	−22.06	8.50	1642.16	0.029
Al-H-10%, + Cat-4%	5.81	50.21	−38.49	15.71	28.15	23.53	−22.03	8.02	1588.25	0.029
Al-H-15%, + Cat-4%	5.37	54.12	−42.59	16.86	28.82	24.75	−22.41	9.40	1667.54	0.029
Al-H-20% +, Cat-4%	5.21	51.96	−39.72	17.33	28.66	24.91	−22.38	7.18	1610.89	0.029
Al-H-5% + Cat-6%	5.99	57.38	−45.91	17.94	29.37	25.85	−22.73	8.60	1723.81	0.029
Al-H-10%, + Cat-6%	**6.98**	**55.12**	**−42.88**	**16.84**	**30.27**	**25.90**	**−22.98**	**5.62**	**1675.94**	**0.029**
Al-H-15%, + Cat-6%	6.55	54.13	−41.37	16.34	30.12	24.7	−22.28	8.34	1645.32	0.029
Al-H-20% +,Cat-6%	6. 78	52.29	−41.98	17.29	29.67	25.12	−21.29	7.56	1610.12	0.027
Al-H-10%, + Cat-8%	**6.58**	**54.12**	**−41.88**	**15.84**	**29.27**	**24.90**	**−21.98**	**5.92**	**1680.94**	**0.030**

Table 11.
Variation in single and double mordants and dye concentrations caused by the addition of various concentrations of potash alum (Al) plus Harda (H) [Al-: H: 50:50] dual mordant in sequence and 2–8% purified catechin extract as a natural dye.

MLR of 1:30 and a salt concentration of 5 gpl, a 10% overall application of Alum plus Harda (50:50) double mordant was applied in sequence, followed by dyeing with a 6% aqueous extract of purified catechu powder under the abovementioned standardized dyeing conditions.

The above data in **Table 12** indicate a greater dye yield at an acidic pH of 4–5 than at an alkaline pH of 8–12 for 10% overall Al + Harda (50:50) dual premorded cotton, which had the highest k/s value of approximately 6.95–6.98.

3.6 Analysis of color fastness to washing, light and rubbing for the use of single and double mordants

The wash fastness of the natural dyed cotton fabrics was evaluated according to the IS 3361–1979 method. The lightfastness of the naturally dyed cotton fabric was evaluated according to the IS: 2454: 1985 method. Additionally, the rubbing fastness of naturally colored cotton textiles was evaluated by the IS M766–1988 method [23].

Thus, the relevant data in **Table 13** indicate that dual mordanting provides better color yield and higher wash fastness.

Sample	K/S	ΔE	ΔL	Δa	Δb	ΔC	ΔH	BI	MI	CDI
Control	0.055							80.12		
H-10%- Cat-2%	3.28	50.45	−36.96	17.09	29.77	25.66	−22.81	9.63	1556.37	0.028
H-10%-Cat-4%	3.98	49.45	−36.06	16.19	28.97	24.76	−22.20	9.33	1520.37	0.030
H-10%-Cat-6%	4.77	48.38	−36.11	15.61	28.16	23.48	−22.03	9.15	1516.19	0.029
Al −10%-Cat-2%	4.57	49.97	−36.83	17.87	28.67	25.25	−22.44	6.74	1548.97	0.027
Al −10%- Cat-4%	4.81	50.21	−38.49	15.71	28.15	23.53	−22.03	6.02	1588.25	0.029
Al −10% - Cat-6%	4.98	56.12	−40.88	15.84	29.27	23.90	−22.08	5.98	1665.94	0.028
Al-H-5%, + Cat-2%	4.46	48.95	−38.58	15.62	25.76	21.57	−21.02	10.23	1583.57	0.030
Al-H-10% +,Cat-2%	4.57	49.97	−36.83	17.87	28.67	25.25	−22.44	10.24	1548.97	0.027
Al-H-15% +,Cat-2%	5.19	50.94	−39.49	16.66	27.51	23.59	−21.36	10.27	1604.94	0.028
Al-H-20% +,Cat-2%	5.46	51.73	−41.53	16.51	26.05	22.36	−21.24	9.16	1641.49	0.029
Al-H-5%, + Cat-4%	5.78	52.73	−41.37	17.00	27.92	24.14	−22.06	8.50	1642.16	0.029
Al-H-10%, + Cat-4%	5.81	50.21	−38.49	15.71	28.15	23.53	−22.03	8.02	1588.25	0.029
Al-H-15%, + Cat-4%	5.37	54.12	−42.59	16.86	28.82	24.75	−22.41	9.40	1667.54	0.029
Al-H-20% +, Cat-4%	5.21	51.96	−39.72	17.33	28.66	24.91	−22.38	7.18	1610.89	0.029
Al-H-5% + Cat-6%	5.99	57.38	−45.91	17.94	29.37	25.85	−22.73	8.60	1723.81	0.029
Al-H-10%, + Cat-6%	**6.98**	**55.12**	**−42.88**	**16.84**	**30.27**	**25.90**	**−22.98**	**5.62**	**1675.94**	**0.029**
Al-H-15%, + Cat-6%	6.55	54.13	−41.37	16.34	30.12	24.7	−22.28	8.34	1645.32	0.029
Al-H-20% +,Cat-6%	6. 78	52.29	−41.98	17.29	29.67	25.12	−21.29	7.56	1610.12	0.027
Al-H-10%, + Cat-8%	**6.58**	**54.12**	**−41.88**	**15.84**	**29.27**	**24.90**	**−21.98**	**5.92**	**1680.94**	**0.030**

Table 12.
Effects of variation of single and double mordants and dye concentrations by varying concentration of potash alum (Al) plus Harda(H) [Al-: H: 50:50] dual mordant applied in sequence and 2–8% purified catechu extract as natural dye.

Moreover, by changing the extraction solvent or varying the extraction conditions, the color depth may be altered in the case of natural dyes applied to any textile. To eliminate such variations, separate standardization of dyeing process variables with the use of different mordant combinations (for example, the use of K-Al and Gall nut combinations instead of K-AL + Harda combinations results in different color yields). Moreover, alcoholic extraction, sonicator-assisted extraction or microwave-assisted extraction results in better color yields and other responses, such as UV resistance or antimicrobial properties, at different scales.

Thus, from the results of varying concentrations of single and double mordants and varying dyeing process variables, etc., and color fastness to wash, light and rubbing, as shown in **Tables 11–13**, few important points of observations arise from these concentrations, which may be explained scientifically as follows:

3.6.1 Why acidic pH shows better color yield than alkaline or neutral pH of dye bath?

Initially, upon the immersion of cotton in dye bath water, the surface of cotton contains ions that do not support the absorption of anionic natural dyes; however,

Mordant and Dye Used	Color Fastness to				
	Washing		Light	Rubbing	
	LOD	ST		Dry	Wet
H-10%-Cat-2%	3	3	3	3	2
H-10%-Cat-4%	3–4	3–4	2–3	3–4	1–2
H-10%-Cat-6%	3	3	3	3–4	1–2
Al –10%-Cat-2%	3–4	3–4	3–4	3–4	2–3
Al –10%- Cat-4%	4	3–4	3–4	3–4	2–3
Al-10% - Cat-6%	3–4	3	3	3	2
Al + H-10%(50:50)- Cat-2%	3–4	3–4	3–4	3–4	2–3
Al + H-10%(50:50)- Cat-4%	4	4	3–4	3–4	3
Al + H-10%(50:50)- Cat-6%	4	4	3–4	4	3

Table 13.
Color fastness results for 10% Harda single premordanted and 10% overall concentrations of alum and Harda (50:50) dual premordanted cotton fabric dyed with varying concentrations of catechu.

cationic materials and acid protons gradually negate this negative charge on cotton and allow more dye to be absorbed on premordanted and/or acidic protonated cotton to absorb more anionic natural dyes. Moreover, the acidic pH of the dye bath facilitates the quicker and greater anionization of natural dye anions in the dye bath, allowing more dye anions to be absorbed by the cotton than does the alkaline pH of the dye bath. In addition, the anionic part of the acidic salt/acid mixture used in dye baths causes the common ion effect to make dye absorption slightly slower as a retarder; hence, more uniform dyeing results in an acidic pH bath, which is an added advantage in this case.

3.6.2 Why does the use of double mordants produce a greater color yield than the use of a single mordant?

The simple answer is to use dual or double mordants using two types of mordants that act in two different ways for fiber-mordants and natural dye complexing, where the metallic mordant, such as Al-from alum, acts to form complexes with natural dyes either by producing coordinated covalent bonds or simply by hydrogen bonding or both, while the 2nd mordant harda or gall nut, being tannates of biossources, acts to form larger dye-fiber complexes with these already aluminum mordanted fibers and dyes utilizing their tannate/tannic acid functionally; thus, a larger/larger complex of [cotton fiber mordant 1-mordant-2-natural dye] with two mordants is formed, providing superior dyeability and better color fastness to wash.

3.6.3 Is there any reason why different tannin-based biomordants show different results when dyeing cotton with the same natural dyes?

The simple explanation is that it is obvious that two types/two different biomaordants have different functionalities. For example, harda containing chebulinic acid provides functional COOH and few OH groups to form more complexes with both dye and fibers, while gall nuts containing elagic acid and gallic acid tannates or flavonoids provide much higher tannate contents than tannate content of harda; hence, such complexing of these tannate-based biomordants with fiber and natural dye is expectedly more or more for gall nut extract than harda extract used separately as biomordant.

3.6.4 Is there any differences in extraction of color components under two different solvents, if so, why?

The simple answer is that different types of solvents have different abilities to extract color components and are associated with other contents in the two different extracts obtained by the use of two different solvents or their mixtures.

For example, when eucalyptus leaves or barks are extracted with boiling water or an alcoholic solution, the results of the compositional analysis of these two extracts differ. Water extraction results in yellowish light brown liquor with less eucalyptol and other components, but 50:50 water and ethyl or methyl alcohol-based extracts at 70°C result in much deeper brown liquor with more or greater percentages of colored eucalyptol and other components extracted, as well as few UV-resistant components, which are not extractable by water boiling only.

4. Concluding remarks

Thus, the use of a UV VIS Reflectance Spectrophotometer and a UV VIS Absorbance Spectrophotometer attached to a suitable high-definition Computer-CPU/work station and dedicated software for managing instrumental color measurements, storing a color database and maintaining records of color match prediction and match corrections for textile and apparel products has become common practice in a textile dyeing finishing process house for maintaining color quality by achieving precision match results.

In addition, quality control activities for checking color quality for dyed textiles and apparels, i.e., determination of batch-to-batch color differences after dyeing each new batch, were performed to evaluate differences in color values in terms of the DE*, DL*, Da*, and Db* color difference CIE 1976 scale of color measurements.

Computer-aided UV–VIS reflectance spectrophotometers and suitable application software have been used to modernize dye houses in the textile and apparel industry to determine the quality of the dyeing results.

Thus, dye uniformity and reproducibility of color match results have become easily possible; hence, any industrial colorist can confidently reproduce any color/shade using computerized color measurements and match prediction systems within a permissible minimum total color difference (DE) of DE = 1 for cotton, silk, and woolen textiles and within a DE value = 2 for jute-based textiles/decorative colored/printed bags, home furnishings, etc.

The effects of any variation in the dyeing process parameters or dyeing conditions on the color yield and colourfastness to wash, light and rubbing fastness of any textile fabric dyed with direct, reactive, or natural acid dyes can be studied, and accordingly, e-Shade cards can be prepared. However, in some cases where the color fastness is not adequate, methods for improving the color fastness can be tested via different posttreatment methods.:

Pilot plant trials for the reproduction of computer-aided dyeing recipes/formulations using a Lab mini jigger have been completed, and the accuracy of the database for determining color matching recipes can be checked by using the color matching index as a newer index.

To produce compound shades, testing for compatibility of individual dyes in a binary pair of dye mixtures can also be performed. The quality of dyes, dyeing uniformity, etc., and other important color parameters for the production of wash-fast and lightfast dyed textile and apparel products can be separately identified for the use of precision color quality.

Author details

Ashis Kumar Samanta
Department of Jute and Fiber Technology, Institute of Jute Technology, University of Calcutta, Kolkata, West Bengal, India

*Address all correspondence to: ijtaksamanta@hotmail.com

References

[1] Trotman ER. Dyeing and Chemical Technology of Textile Fibers. 5th ed. London: Charles Griffin & Co Ltd; 1975. pp. 561-562

[2] Shah HS, Gandhi RS. Instrumental Color Measurements & Computer Aided Color Matching for Textiles. India: Mahajan Book Distributers: Ahmedabad; 1990. pp. 76-116

[3] Samanta Pubalina, Chapter 5: Basic principles of color measurement and color matching of textiles and apparels, in a Book on Colorimetry, ed. by A K Samanta (Published by London, UKInTechOpen, London and Croatia, UK), 2022, 105-128

[4] Samanta AK. A review and case study of computer application in jute or textile mills' dye house and computerized color matching. Textile Trend. 1993;**10**:37-47

[5] Samanta AK, Das D. Studies on quantitative color measurement of direct dyed jute fabric in relation to computerized color matching. Journal of Institution of Engineers (India) Textile Engineering. 1992;**73**(12):53-57

[6] Davidson HR, Hemmendinger H, Landry JLR Jr. A system of instrumental color control for the textile industry. Journal of the Society of Dyers and Colorizts. 1963;**79**(577–589):601-603

[7] Park RH, Stearns EI. Spectrophotometric formulations. Journal of the Optical Society of America. 1944;**34**:112-113

[8] Allen E. Basic equations used in computer color matching. Journal of the Optical Society of America. 1966;**59**(9): 1256-1259

[9] Stearns EI. Computer color matching. American Dyestuff Reporter. 1964;**53**: 668-672

[10] Randall D, Stutts T. Optimizing calibration dyeing for computer color matching. Amer Dyestuff Reporter. 1988;**28**:44-46

[11] Davidson HR, Friede E. The size of acceptable color differences. Journal of the Optical Society of America. 1953; **43**(7):581-589

[12] Kuehni R. Some practical improvements in computer shade matching. Color Engineering. 1969;**7**(5): 52-54

[13] Karbasi A, Moradian S, Asiaban S. Improving the performance of computer color matching procedures. Journal of the Optical Society of America. 2008; **25**(9):2251-2262

[14] Samanta AK. Chapter 1: Colorimetric evaluations and characterization of natural and synthetic dyes/pigments and dyed textiles and related products, in a Book on Colorimetry, ed. by A K Samanta (Published by London, UKInTechOpen, London and Croatia, UK). 2022. 3-33

[15] Samanta AK, Agarwal P, Singhee D, Dutta S. Application of single and mixtures of red sandalwood and other natural dyes for dyeing of jute fabric: Studies on color parameters/color fastness and compatibility. Journal of Textile Institute. 2009;**100**(7):565-587

[16] Samanta AK, Agarwal P, Datta S. Studies on color strength related parameters and compatibility for dyeing of cotton fabrics with binary mixtures of jackfruit wood and other natural dyes. Journal of Natural Fibers. 2009;**6**:27-45

[17] Sinnur HD, Verma DK, Samanta AK. Compatibility of binary mixtures of

natural dyes for developing compound shades for cotton khadi fabric. Indian Journal of Fiber & Textile Research. 2021;**46**:158-167

[18] Samanta AK, Konar A, Sekhar BN, Ray AN. Studies on compatibility of selective direct dyes for dyeing of jute fabric. Indian Journal of Fiber & Textile Research. 2019;**44**:98-106

[19] Sinnur HD, Kar TR, Verma DK, Samanta AK. Effect of dyeing process variables on color yield and color fastness properties for cotton khadi fabric dyed with deoiled redsandal wood waste. Journal of Institution of Engineers (India)–Series-E (Chemical and Textile Engineering). 2021;**102**:273-292

[20] Samanta A K & Samanta Pubalina, Eco-friendly approaches for improving color fastness, UV protection and antimicrobial properties with natural resource based functional finishing of natural dyed silk and cotton fabrics" a book chapter in New World of Textiles: Shaping for the Future, ed by K.N. Guruprasad, Ranjit Kr Datta, Tapan Kumar Sinha & Swapnil Jain. Delhi, India: SVITT, SVV, Indore, Publisher Allied Publishers Pvt Ltd; 2021. Ch-5. p. 45-73

[21] Pubalina S. Fundementals and applications of computer aided color match prediction (CCMP) system. Latest Trends in Textile and Fashion Design. 2018;**2**(5):1-10

[22] IS 17084: 2019. Textile Dyestuffs-Natural Indigo-Identification. New Delhi, India: BIS; 2019. pp. 1-18

[23] BIS. Handbook of Textile Testing. 1st ed. New Delhi: Bureau of Indian Standards; 1982, Rev. edn (1986). pp. 536-573

Section 2

Fundamentals and Applications of Colorimetric Analysis

Chapter 2

Fundamentals of Colorimetry

Yagya Kumari Shrestha and Shree Krishna Shrestha

Abstract

There are several kinds of analytical techniques following the principle of photometry in which colorimetry comes under absorption photometry. The colorimetry is commonly used analytical technique involved in quantitative estimation of color i.e. it is utilized to find out concentration of the colored substance in the sample solution e.g. water, biological samples at visible spectrum of light (380–780 nm). The colorimeter is an instrument in which this technique is used. It is also called absorptiometer. A substance must be colorful or should have property of forming chromogens through the addition of reagents which will absorb light according to their color intensity to be measured. The intensity of the color is in proportion to the concentration of colored compound. Most of the analytical techniques used in our clinical laboratory presently are based on this photometric principles in which absorbed, transmitted or emitted light are measured. When intensity at different wavelength on the whole range of electromagnetic spectrum is measured, it is called spectrophotometry. Smartphone accessories have been evolved to allow the simple, quick reproducible values of the molecules.

Keywords: Beer's and Lambert's law, chromogen, colorimetry, photometry, standard curve

1. Introduction

Colorimetry is a type of photometry which is basically considered as the techniques in which light is detected and also detects changes in its intensity. The root word "photo," means light. A photometer is a machine measuring the energy of electromagnetic wave in the range of infrared radiation to ultra-violet radiation, including the visible part of the electromagnetic spectrum (**Figure 1**). It changes light into electric current by using a photocell. Colorimetry is referred if the measured light is in the visible range of the electromagnetic radiation. In this method a beam of light from a light source is allowed to pass via sample holder containing the analyte in the solution, the intensity of light transmitted will be less than the light passing through sample in the cuvette. The absorbed light is in proportion to the concentration of the analyte. The color of the sample is an intrinsic characteristic of the solution or it can be evolved by the addition of suitable reagents. The absorption of the sample is compared to that of standards from which the concentration of test sample can be calculated [1–3].

Photometric methods can be divided into two catagories, visual and physical. The physical photometry is commonly practiced.

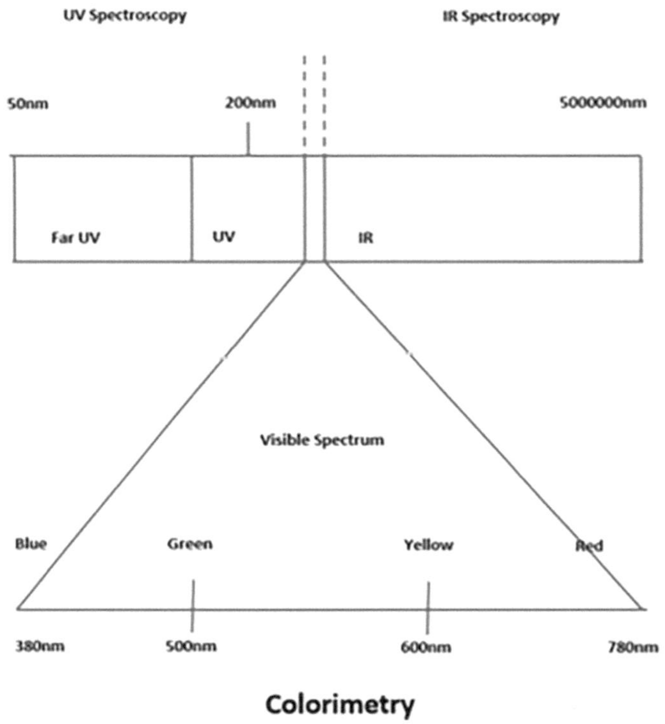

Figure 1.
Visible electromagnetic spectrum.

There are various kinds of analytical techniques that are based upon photometric principles. In Spectrophotometry, atomic absorption, turbidometry etc. absorbed or transmitted light is measured while in flame emission photometry emitted light is measured.

2. History

Scientist James Clerk Maxwell from Scotland, invented colorbox in 1860 which was the primitive colorimeter featuring a prism that assisted in control of red, green, and blue light beams separately to match the color in the sample. Later Louis Jules Duboscq (1817–86), a French optical instrument maker, designed Duboscq colorimeter in 1854 (**Figure 2**). It was one of the early colorimeters to be one of the popular. It allowed for the coexistent color comparison of two liquids. Duboscq explained an improvised variety to the French Academy of Sciences in 1868. A Duboscq colorimeter calculates the concentration of a substance through a visual comparison of the color intensity of the compound against that of a standard solution. This method of identification was revolutionary when first introduced, but the colorimeter was replaced by the emergence of the more precise spectrophotometer in the early 1940s. Bausch & Lomb then remodeled the first Duboscq colorimeter (**Figure 3**).

Later photoelectric colorimetry became widely known in the mid-1930s, and in 1938 William Henry Summerson brought up colorimeter containing a photocell. There is fascinating background on use of photocells which were mainly used by Germans. After the world war was stopped, an English pupil Arthur Evans was acquainted to a

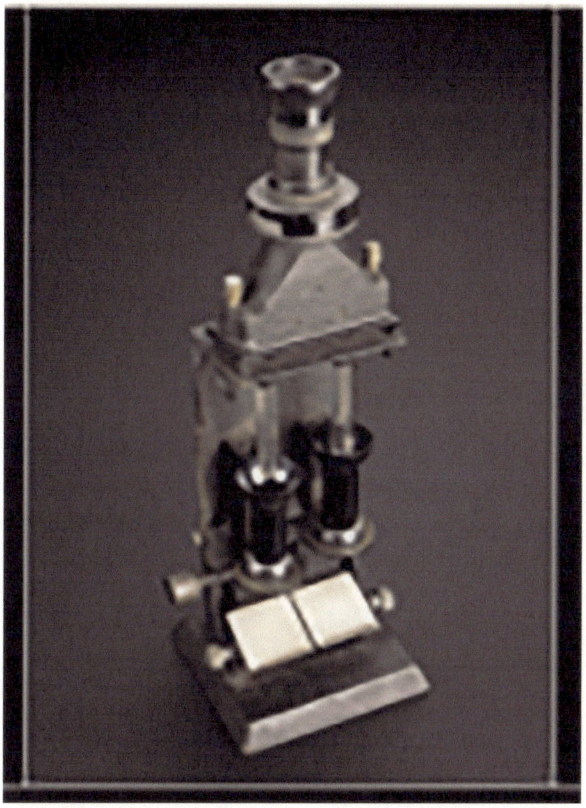

Figure 2.
Duboscq colorimeter, by F. Hellige and Co., Freiburg, Germany, 1901–1914.

leading medical researcher Dr. Rose of the Hammersmith Hospital, (UK) was looking for inventing a device which would enable him calculating the intensity of a color in a solution thereby calculating the concentration of an element in it. Colorimetry was already in use widely as analytical technique in medical, scientific and industrial laboratories which was performed by matching the solution under direct visual examination to known standards. Evan's EEL company started to produce first selenium photocell colorimeters (the model 2) commercially and were distributed all over the world. It gave both accurate and reliable results. Later the company was bought by Corning which was the pioneer company to produce pH sensitive glass for pH electrodes which are still popular in medical and industrial laboratories. However, new colorimeters do not use selenium photocells. This pioneer company is still gaining it's momentum in this field.

Later Klett Bio Colorimeters were made for clinical laboratory work. This Klett colorimeter includes standard color disks made of glass, which can be used in place of standard solutions in color comparisons [4–6].

2.1 Terminologies

Light: It is an electromagnetic radiation (EMR). It is composed of γ rays, X rays, ultra violet rays, visible rays, infrared rays, radio waves and micro waves. Of the components only those visible to eye is known as white light [7–10].

Figure 3.
Bausch & Lomb Duboscq Colorimeter.

Photons: Light is energy which exists in the form of bundles or as discrete packets of quanta called photons. The energy (E) of EMR is directly proportional to its frequency (u) and velocity (c) but inversely proportional to its wavelength (λ).

Thus, $E \alpha u$But, $u = c/\lambda$So, $E \alpha c/\lambda E = hu = h c/\lambda$

where, h = plank's constant, c is the velocity of light in vacuum, 3×10^8 m/sec.

EMR is produced by events at the molecular, atomic, or nuclear level. Some of the events which give rise to EMR are oscillations of nuclei and electrons in electrical or magnetic fields, molecular bending and vibrations, excitation of orbital electrons, ejection of an inner orbital electron and rearrangement of the other electrons, nuclear break-up etc. The radiation they emit will have different wavelengths as each of these events differ in terms of the energy involved (**Table 1**). Thus, a full spectrum of EMR will be produced. Their magnitude of electrical vector and magnetic vector are denoted by the symbol **E and H** respectively. A beam of light from a light source consists of many randomly oriented plane polarized components being propagated in the same direction (**Figure 4**).

Polychromatic light: A beam of waves from a light source consists of several wavelength and is known as polychromatic.

Monochromatic light: A beam of waves in which all the rays have the same wavelength is known as monochromatic.

Spectrum: When EMR allowed to pass through a prism, all the 7 rays are dispersed into an orderly pattern called spectrum and can be measured by a spectrometer. The components have different wavelength and frequency. It can be classified as emission spectrum & absorption spectrum. When radiation is directly examined by a spectrometer the resultant pattern of wavelength is called emission spectrum. If a radiation is examined after interacting with an absorbing medium it is called absorption spectrum. Emission and absorption spectrum may be either line or band type. Spectrum produced by atoms in gaseous state appears as lines and it is called line or atomic spectrum. Spectrum produced by molecules in solution or crystal appears as bands

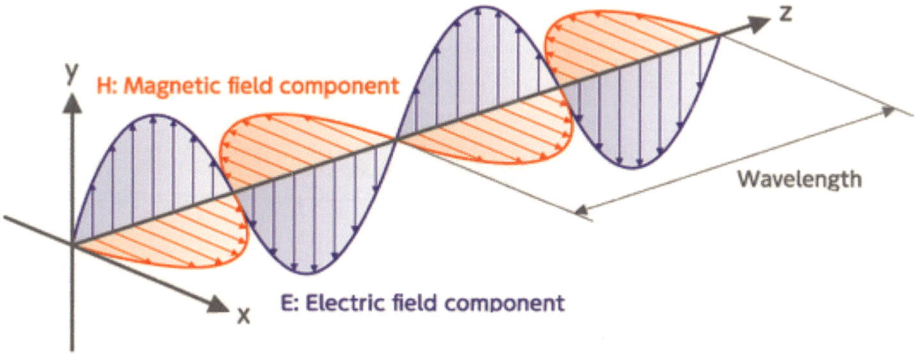

Figure 4.
Electromagnetic radiation wave of which is propagating transversally in space and time consists of the electric (E) and magnetic (H) field vectors directed perpendicular to each other.

Region	X-rays	Ultraviolet	Visible	Infrared	Microwave
Wavelength	0.1–100 nm	100–380 nm	380–750 nm	750 nm to 100 μm	100 μm to 30 cm
Effect on molecules	Excitation of sub-valence electrons	Excitation of valence electrons	Excitation of valence electrons	Molecular vibrations	Molecular rotations

Table 1.
Effect of absorption of electromagnetic radiation on the molecules.

and it is referred as band or molecular spectrum. Emission spectrum are of line & band emission spectrum while Absorption spectrum are of line & band absorption spectrum.

Spectroscopy: It is a technique which is used for quantification, characterization and structural analysis of a substance based on interaction of light with that substance. Interaction of light with a substance may result in different phenomena and generate a specific analytical signal. The signal is detected by a detector and recorded by a digital recorder in the form of a spectrum. The spectrum can be correlated with the amount, structure, property and functional group of the sample.

a. Emission spectrum

It is characteristic of the emitting atom or molecule in excited state. Line emission spectrum appears as fine bright lines on a dark background. e.g. line emission spectrum of sodium atoms excited by flame. Band emission spectrum appears as broad bright bands. e.g. Band emission spectrum of Ca salts heated in a flame.

b. Absorption spectrum

It is characteristic of the emitting atom or molecule in ground state.
Line absorption spectrum appears as dark lines on a bright background e.g. line absorption spectrum of sodium atoms irradiated by white light. Band absorption spectrum appears as broad dark bands e.g. Band absorption spectrum of iodine vapor irradiated with white light (**Figure 5**).

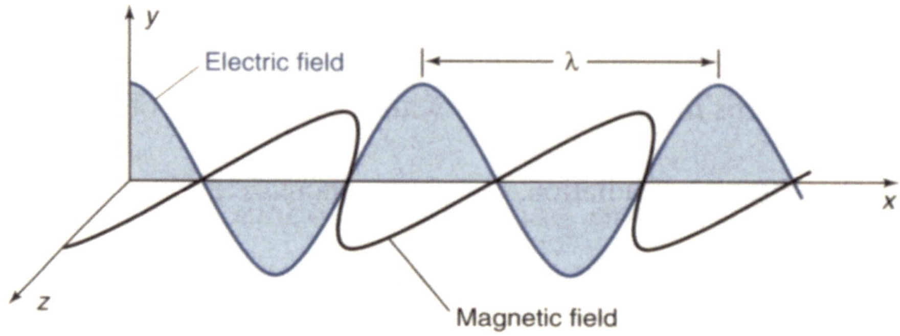

Figure 5.
EMR spectrum.

Frequency: When radiation is propagated into a medium, it oscillates. Number of oscillations occurring per second is called as the frequency (**u**) of the radiation. It is expressed as Hertz (**Hz**).

Wave number: Number of waves propagating per centimeter of the medium is called its wave number (**σ**) or **ū**.

Wavelength: The distance between two successive peaks of a wave is known as wavelength (**λ**) (**Figure 5**). The wavelength of a light wave determine its color or visibility. For e.g. light wave with λ between 380 and 780 nm is visible to human eye and therefore called visible light or white light. Visible light has a range of colors which are abbreviated in mnemonic VIBGYOR.

Velocity: Distance traveled by the light wave per second is called its velocity (**c**). It is expressed as cm/sec.

3. Parts of colorimeter

The parts of the colorimetry are as follows [3, 7, 9–11]:

1. **Light Source/Radiation source:** An electric lamp is used as radiation source mostly. Commonly used lamp is tungsten filament lamp (for wavelength 320–700 nm). Any lamp source which gives adequate intensity of radiation over the entire wavelength region can be used. The source of light for colorimetry should provide continuous radiation from 320 to 700 nm with adequate intensity which should be stable and fluctuations free. As the tungsten filament lamp fulfills the above criterias, this lamp makes its place in the most of colorimeters. The major drawback of the tungsten lamp is that it mostly emits its radiant energy in the near-infrared region of the spectrum. But at higher temperatures the life of the lamp is much reduced. In order to remove the unwanted infrared radiation a filter that absorbs heat is placed between the lamp and sample which absorbs most of the infrared radiation without affecting radiation energy at other wavelengths. Only about 15% of radiant energy emitted lies in the visible region.

2. **Slit:** It allows only a beam of light within visible spectrum excluding unwanted light.

3. **Condensing lens**: A beam of light after passing through slit falls on condenser lens which turns into parallel beam of light.

4. **Filters or monochromator:** Only light of required wavelength is passed through filter or monochromator while light of other wavelengths are absorbed.

 Filter: It permits only monochromatic light to pass through while absorbs unwanted ones. It is usually made up of colored glass or dyed gelatin. This is a means of selecting light of narrow wavelength λ (50 nm or more). A green filter allows green color to pass through while rest are absorbed. The color of filter used is always complementary to color of solution.

 I. Absorption filter: It may be of glass absorbance or gelatin absorbance type. Filter is solid sheet of glass being colored by a pigment (dyed gelatin or similar materials) which is dissolved or dispersed in the glass.

 II. Interference filter: It may be of constructive interference or destructive interference (Fabry-Perot). It has dielectric spacer film made up of CaF2, MgF2 or SiO, between two parallel reflecting silver films. The thickness of dielectric spacer film can be 1/2λ (1st order), 2 λ/2 (2nd order), 3 λ/2 (3rd order) etc.

 Monochromators: These are better and more efficient than filters in converting a polychromatic light into monochromatic light. It is usually found in spectrophotometer. Monochromators are of various types:

 I. Prism: It is either of refractive type prism or reflective type prism.

 II. Grating: It may be diffraction grating or transmission grating.

5. **Cuvette/sample cell/sample holder:** They are specially designed tubes made up of optical glass (borosilicate or quartz) to hold the colored sample for measurement in colorimeter for accurate reading. There are square, rectangular or round shaped cuvettes. It is used to hold samples of fixed optical pathlength which is usually 1 cm. Its capacity is 3–4 ml or even lesser in case of micro cuvettes. The colored solution in cuvette absorbs the complementary color (**Table 2**). It is made of material that does not absorb light of wavelength range of the interest.

Color of solution	Filter	Wavelength (nm)	Range (nm)
Blue green	Red	620	650–700
Blue	Yellow	590	570–600
Purple	Green	550	505–555
Red	Blue green	500	495–505
Orange	Green blue	490	475–495
Yellow	Blue	445	420–475
Yellowish green	Violet	410	400–420

Table 2.
Colors of solutions and their complementary filters.

Color wheel: The Color of Light *absorbed* and *observed* passing through the compound are complementary. Complementary colors lie across the diameter on the color wheel and combine to form "white light", so the color of a compound seen by the eye is the complement of the color of light absorbed by a colored compound.

6. **Detector (photocell)**: A device used to convert the light to electrical energy, comprising metal plate coated with photosensitive element such as selenium or cadmium (+ve electrode). The light sensitive material is covered with a transparent thin layer of gold or copper (−ve electrode). When the light strikes the selenium layer, electrons are liberated which pass into the transparent layer making it electronegative. A potential difference is created between the transparent layer and metal plate. The electric current thus generated is directly proportional to the intensity of light striking the photocell.

7. **Readout device**: The potential difference or electrical signal in photocell is detected by readout device. The device also can be called galvanometer or ammeter or digital readout, which shows absorbance & transmittance or both.

4. Principle of colorimetry

Principle of colorimetry: When a beam of monochromatic light passes through a colored solution, the coloring substances absorbs a portion of the light & the rest is transmitted. Absorption of light is related to the color intensity. The color intensity will be proportional to the concentration of the chemical (analyte) responsible for producing the color [7–11].

Absorbance (A): When light passes through a medium, the ratio of log of intensity of incident light to the intensity of transmitted light is called as absorbance or optical density (OD). Previously it was called extinction.

Transmittance (%T): When light passes through a medium, the ratio of intensity of the transmitted light to the intensity of incident light is called as transmittance.

Path length (l): The internal cross length of the cuvette through which light passes is called as path length. Usually it is 1 cm.

Absorbance maxima: Wavelength of maximum absorbance is known as absorbance maxima. It is denoted by λ_{max}. The wavelength at which a substance shows maximum absorbance is called absorption maximum or λ_{max} (**Figure 6**).

The amount of light is absorbed or transmitted by a colored solution in accordance with 2 laws.

1. **Beer's Law:** When monochromatic light passes through a colored solution, amount of light transmitted decreases exponentially with increase in concentration of colored substance. i.e. the amount of light absorbed by a colored solution is directly proportional to the concentration of substance in the colored solution **A∝c** (**Figure 7**).

2. **Lambert's law:** The amount of the light transmitted decreases, exponentially with increase in pathlength (l) of the cuvette or thickness of colored solution through which light passes. i.e. the amount of the light absorbed by a colored

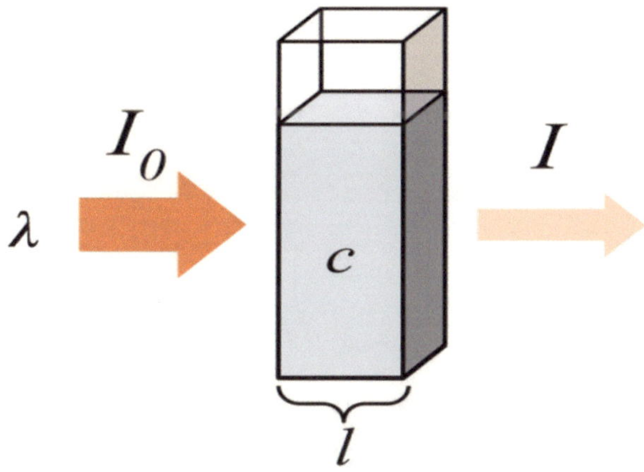

Figure 6.
Showing relationship between absorbance, absorptivity, pathlength and concentration where, io = incident light, I = transmitted light, a = absorbance, c = concentration of absorbing compound, l = pathlength, ε = molar absorptivity.

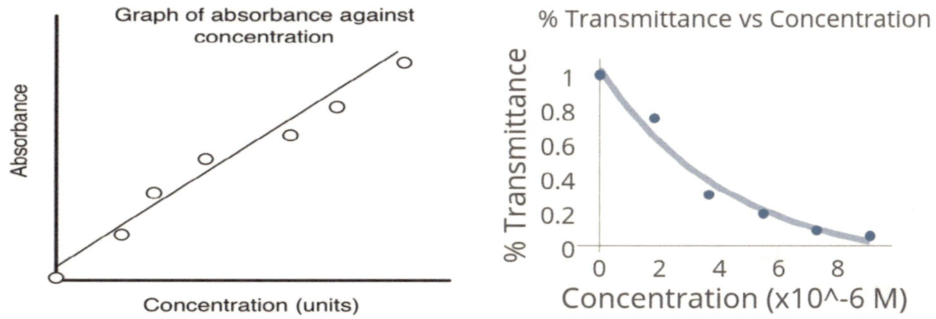

Figure 7.
Beer's law.

solution depends on pathlength of cuvette or thickness or depth of colored solution **Aαl** (**Figures 8** and **9**).

Relationship between absorbance (A) and transmittance (T): Relationship between (A) & (T) i.e.

$Io = A + I$ $A = Io–I$.

By definition, $A = \log Io/I$.

$A = -\log I/Io$.

$A = -\log T$ (As $T = I/Io$).

$A = \log 1/T$.

$A = \log 100/(\%T)$ to take into percentage multiply by 100.

$A = \log 100 – \log\%T$ (As $\log a/b = \text{Log } a – \text{Log } b$).

$A = 2 - \log\%T$ (As $\log 100 = 2$).

Combined Beer's – Lambert's law: It is denoted as, amount of light transmitted through a colored solution decreases exponentially with increase in concentration of colored solution & increase in the pathlength of cuvette or thickness of the colored solution.

$$\mathbf{A \alpha c} \text{ or } \mathbf{A = k\ c} \tag{1}$$

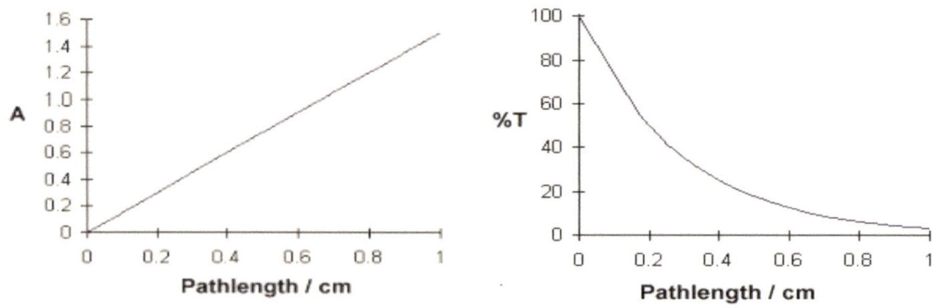

Figure 8.
Lambert's law.

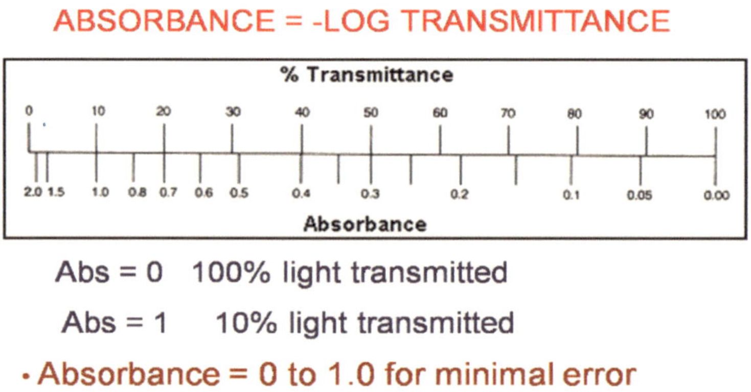

Figure 9.
Relationship between absorbance (a) and transmittance (T).

Where, k = linear absorption coefficient of the absorbing material.

$$\mathbf{A \alpha l \; A = k' l} \tag{2}$$

Where, k′ = absorptivity constant.

Combinely from Beer-Lambert law both k & k′ can be merged to form a single constant a & hence, combined equation can be expressed as **A = acl (Figure 6)**

Where,

A = absorbance.

a = proportionality constant defined as absorptivity.

c = concentration of the absorbing compound sample in solution usually expressed in grams per liter.

l = pathlength of the sample (cm).

A = acl, this equation forms the basis of quantitative analysis by absorption photometry. **A** values have no units, hence, the units for **a** are the reciprocal of those for **l**, when **l** is **1 cm** and **c** is expressed in moles per liter the symbol ε (epsilon) is substituted for the constand **a**.

A = εcl,

where,

ε = molar absorptivity/ molar absorption coefficient/molar extinction coefficient ($Lmol^{-1}$ cm^{-1}).

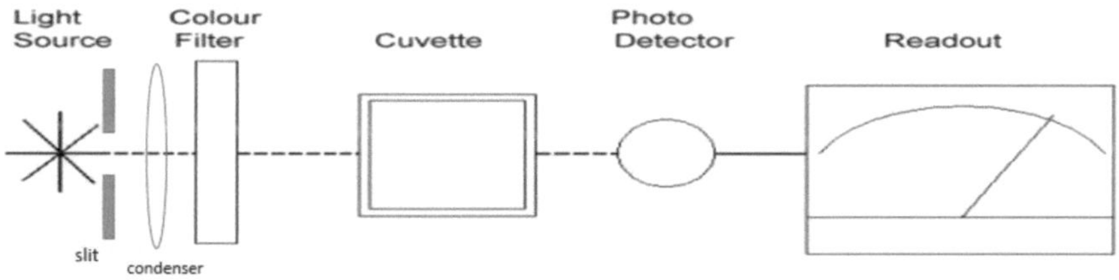

Figure 10.
Working principle of colorimetry.

The value for molar absorptivity ε is constant for a given compound at a given wavelength under prescribed conditions of solvent, temperature, pH, etc. It is a measurement of how strongly a chemical substance absorbs light at a given wavelength. For any particular molecular type, absorptivity changes as wavelength of radiation changes.

Let,

A_T = Absorbance of test A_S = Absorbance of standard.

C_S = Concentration of standard C_T = concentration of test.

$A_T = k \text{ x } C_T \text{ x } L$ and $A_S = k' \text{ x } C_S \text{ x } L$.

$A_T/A_S = (k \text{ x } C_T \text{ x } L)/(k' \text{ x } C_S \text{ x } L)$.

$A_T/A_S = C_T/C_S)$.

Therefore,

$\mathbf{C_T = A_T/A_{S\times}\ C_S}$

i.e. Concentration of test = Absorbance of the test X concentration of a standard/ Absorbance of the standard.

Therefore, above equation states that optical density (OD) is proportional to the sample concentration if the pathlength is constant. It follows then, that if the OD of a standard solution is known, then an unknown sample concentration can be calculated from its OD by applying the above formula. The standard chosen should have an optical density nearer to the OD of the unknown. The most accurate way of measuring the concentration in the unknown sample is through preparation of a calibrated curve or standard graph from a number of standards (**Figure 10**).

4.1 Preparation of standard graph and evaluation of unknown

A calibration curve is prepared by plotting known concentrations of a given substance against absorbance at a particular wavelength (usually the $\lambda_{max)}$. The relationship between absorbance and concentration is linear if the Beer's law is obeyed (**Figure 11**).

$y = mx$.

Where, y = A, x = concentration and m = ab.

Unknown concentration of a substance is measured by comparing the absorbance of unknown to the absorbance of known concentrations of the same substance measured under exactly identical conditions (same pathlength l usually 1 cm, same wavelength or the same band of wavelength, same temperature, same solvent and same instrument etc).

A plot of absorbance values against known concentrations of a substance is known as a standard graph or a calibration curve and standard curve should always be linear. The linearity of the curve represents that it follows Beer's-Lambert law.

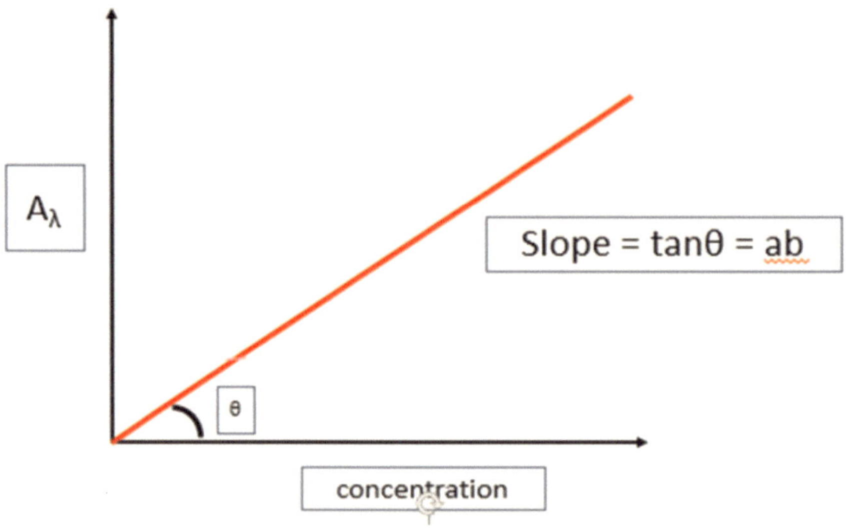

Figure 11.
Preparation of standard graph and evaluation of unknown.

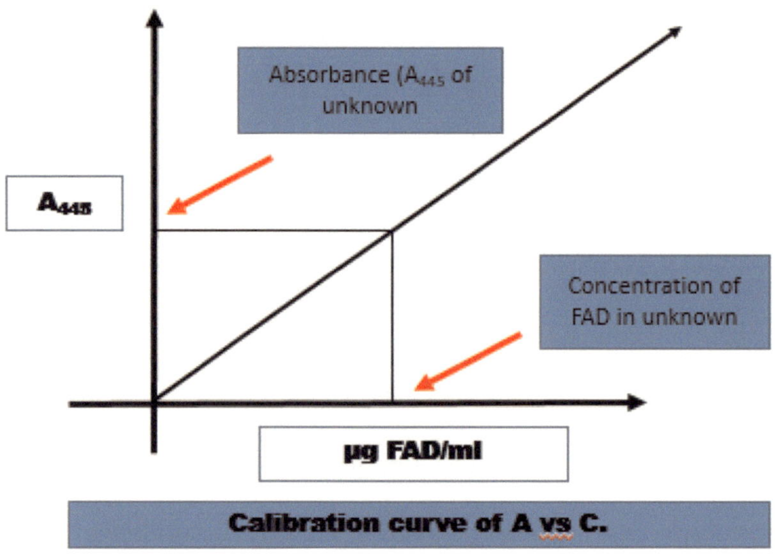

Figure 12.
Evaluation of unknown concentration of FAD from standard graph.

From this graph the unknown concentration of flavin adenine dinucleotide (FAD) can be read off (**Figure 12**).

5. Linearity limit

It is the range of concentrations of an analyte which the instrument can measure linearly beyond which the instrument does not response definetly.

Importance of knowing linearity limit: In biochemistry and other allied sciences, the linearity limit of each test is to be known. It is different for different kits of same parameter. Reading beyond this range is not reliable. In such condition, the sample is diluted and retested and multiplied by the dilution factor. In conclusion calibration curve or standard graph can be used to detect limit of linearity of a test.

6. How to draw standard curve?

Standard curves are defined as a graphs with absorbance or % transmittance plotted on the y axis, and increasing concentrations of standard along the X axis. If the Beer's law is followed, the resulting line representing absorbance vs. concentration will be straight.

A standard curve is constructed after obtaining the (%T) or (A) readings from a number of solutions of known concentration used in a reaction or procedure.

For example, Let us draw standard curve of glucose:

The glucose can be estimated even by plotting graph. This method is helpful to learn more about pipetting and plotting a graph with different concentration of glucose ranging from 50 to 400 mg.

1. Standard glucose (stock) 500 mg of glucose is dissolved in 100 ml of water. It contains 5.0 mg of glucose/ml

2. Working standards: series of working standards of various concentrations are prepared which ranges from 50 to 400 mg/dl.

3. 50 mg/dl – 1.0 ml stock +9.0 ml water

4. 100 mg/dl – 2.0 ml stock +8.0 ml water

5. 150 mg/dl – 3.0 ml stock +7.0 ml water

6. 200 mg/dl – 4.0 ml stock +6.0 ml water

7. 250 mg/dl – 5.0 ml stock +5.0 ml water

8. 300 mg/dl – 6.0 ml stock +4.0 ml water

9. 350 mg/dl – 7.0 ml stock +3.0 ml water

10. 400 mg/dl – 8.0 ml stock +2.0 ml water

Followed by measurement of the absorbance of those standards in colorimeter.

After the readings are obtained each is plotted on semi-log graph paper (% transmittance) or linear graph paper (absorbance) against the corresponding concentration. If the procedure follows Beer's Law, the points plotted will generally lie such that a straight line can be drawn through them. The concentration of controls and other unknowns (patient samples) can be determined by locating their %T or A reading on the line, then dropping an imaginary line down from that point to intersect the concentration axis.

Once a standard curve is developed for a particular test method on a particular spectrophotometer, it should be checked periodically to determine that it is still good.

A new curve should be constructed when there is a change in reagent lot numbers, methodology or procedure, an instrument parameter (change bulb, optics cleaned, etc.),

Once the curve is drawn, a number of things must be considered to determine its acceptability. The majority of the curve's points should be on or close to the line. There could be many reasons for a point not being on the line. Whether or not the curve passes through the point of origin (the "0"), varies with the procedure. If Beer's law is followed and the procedure is linear at the lower concentrations, the curve's line generally goes through the zero.

Always a standard curve should be prepared to confirm that Beer's law is applied to the analysis being carried out. The curve should be linear expectantly. There are several factors that may cause deviations from linearity which refers to deviations from Beer's law. Following factors may cause deviations from Beer's law:

Imperfect monochromacy or polychromatic radiation: The most common factor for deviation from linearity in most colorimeters is the use of a band of wavelengths to measure absorbance. Beer-lambert's law is applicable for monochromatic radiation only.

Other factors which includes followings:

High concentration of chromophores: According to the Beer-Lambert law absorbance (A) is linearly proportional to the concentration of chromophores. There may be deviation in samples with high sample concentrations (>10 mMol of solute) & high absorbance. The intensity of light at choosen wavelength (I_{λ}) should be 10 times higher than the intensity of the stray light (I_{stray}) to reduce deviation from linearity [2]. In addition, the chromophore molecules may dimerize at high concentration and the absorption spectra of the dimers are not the same as that of the monomers. If the spectra differ, the absorption coefficient will differ leading to a positive or negative deviation (**Figure 13**).

Polymerization: It results into the spectral shift. This wavelength at which molar absorption coefficient is not changed is called the isobestic point.

Aggregation: Moreover, high concentration may also lead to aggregation. The intensity of the radiation reaching the detector is decreased because large aggregates scatter light. In such a case a positive deviation will be seen. Aggregation can also lead to electronic interactions that can either decrease or increase the absorption coefficient. At high concentrations, chemical events (association, polymerization,

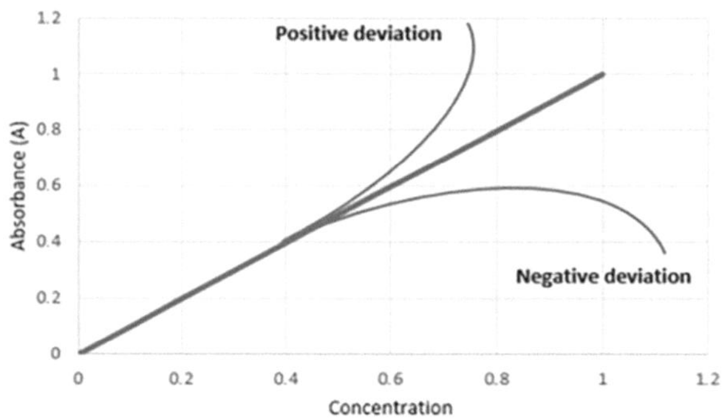

Figure 13.
Deviation from Beer-Lambert law.

dissociation, pH change, interaction with the solvent to produce a product with different absorption characteristics) which may lead to a change in the chemical composition of the solution. Consequently, a deviation from linearity will result.

Low concentrations of chromophores: Deviation may also occur at low concentrations. Proteins are known to denature at low concentrations and, the denatured product has an absorption spectrum that is different from the native protein.

Apart from imperfect monochromacy, other instrumental drawbacks may also result in deviations from Beer's law. Stray radiation reaching the detector changes sensitivity of the detector and power fluctuations of the radiation source and detector amplification system also cause deviation.

Besides above factors, the following factors can also cause deviations from the Beer's law:

1. **Temperature:** There are changes in the degree of solubility, dissociation/association properties of the solute, and several other factors with the changes in temperature. The absorbance is also changed.

2. **Sample Instability:** Some colored compounds are unstable and undergo changes within short duration. In such cases the color might increase, or decrease.

3. **Fluorescence:** Those substances which fluoresce, there might be deviations because fluorescent intensity also reaches the detector apart from the transmitted intensity.

4. **Turbidity:** Turbid solutions give higher absorbance.

5. **Mismatched cell:** If length of sample and blank cell are not equal there may be deviation.

6. **Spectral slit width (SSW):** The deviation occurs if the width of the instrument is not proper. Spectral slit width (SSW) should be as narrow as possible. As SSW increases resolution decreases.

7. Unstable radiation source

8. **Miscellaneous factors**: Impurities, Solvent, pH, suspension, coagulation and so on also deviate from Beer's Lambert law.

7. Variety of applications

- Quantitative estimation of hemoglobin (Hb) and biochemical concentration of analyte in serum sample (glucose, protein, albumin, triglyceride, cholesterol, bilirubin, urea, creatinine, uric acids, calcium etc).

- Quantitative estimation of enzymes such as alkaline phosphatase, alanine transaminase, alpha amylase.

- Testing the water quality for chemicals such as chloride, fluoride, zinc, iron etc.

- Determination of concentration of plant nutrients such as nitrates, NH3.
- Phosphate determination.
- Chrominium in aluminum alloys

In serum sample: colorimetry may be applied for

1. Estimation of serum glucose by glucose oxidase-peroxidase (GOD-POD) method.
2. Estimation of total protein in the given serum sample by Biuret method.
3. Estimation of serum albumin in the given sample by Bromocresol Green Method.
4. Estimation of triacylglycerol in given serum sample by enzymatic method.
5. Estimation of cholesterol in given serum sample by CHOD-4AP method.
6. Estimation of creatinine in given serum sample by Jaffe's alkaline picrate method
7. Estimation of serum bilirubin by modified Jendrassik/Grof method.
8. Estimation of serum calcium based on colorimetric method using (O-cresolpthalein complexone)
9. Estimation of serum urea by enzymatic (urease) method using Modified Berthelot reaction.
10. Estimation of serum uric acid by enzyme urease method (4-AP method).
11. Estimation of alkaline phosphatase in the given serum sample by kinetic method.
12. Estimation of alpha-amylase in the given serum sample by the CNPG3 method.
13. Estimation of alanine transaminase in the given serum sample by the kinetic method.

In whole blood:

1. Hemoglobin estimation by Drabkin method.

In urine:

1. Determination of primaquine metabolites in urine.

2. Determination of urinary chloride.

3. Determination of urinary cystine

4. Quantitative determination of urinary p-aminophenol

In CSF: colorometry is applied for

1. protein determination in CSF.

2. chloride determination in CSF.

3. glucose determination in CSF.

In water:

1. Iron

2. Chloride

Plant material:

1. Phosphorus estimation

8. Pre-requisitive for good colorimetry

1. Steady power supply with voltage stabilizer

2. Centrifuge should be away from colorimeter

3. No bubbles

4. No contamination with thumb prints, scratches

9. Advantages and disadvantages of colorimetry

Advantages: Colorimetry is cheap and useful for quantitative determination of colored substances. It is portable. Since it has limited manual operations it is easy to handle.

Disadvantages: It can not be used for those substances which do not possess color. Errors in inferences can be seen from alike color of interfering compounds. Moreover, very elevated concentrations cannot be measured hence is less sensitive. The filters available are of only narrow range. Unstable light source has possibility of errors since there is change in the light intensities.

10. Case study

- Under aseptic condition total 2 ml of blood from every participant was drawn which was placed in Ethylenediaminetetraacetic acid (EDTA) test tube at school and transpoted to Biochemistry laboratory of Nepal Medical College and Teaching Hospital (NMCTH) in cold blood box and hemoglobin was estimated on the same day [12].

- Hemoglobin (Hb) estimation by Drabkin method was performed at Biochemistry laboratory.

Principle: Hemoglobin (Hb) + Ferriccyanide → Methhaemoglobin → Cyanomethhaemoglobin (stable red compound) which is measured colorimetrically, it's intensity is directly in proportion to amount of Hb.

- **Drabkin method sample:** whole blood

- At biochemistry lab, out of 2 ml whole blood in EDTA test tube, 20 μL (i.e. 0.02 ml) was mixed with 2.5 ml of Drabkin reagent i.e. (tests) in glass tubes with the help of micro pipette then was rinsed for several times for proper mixing. The blank tube was used to adjust the colorimeter to zero optical density at 546 nm wavelength. Finally readings of optical density of the tests was taken which is directly proportional to amount of Hb in the tests as stated by Beer's law of absorbance.

- **Results (Table 3):**

Hemoglobin (gm/dl)	Range	Mean ± SD	Median
	6–22.8	12.63 ± 2.91	12.60

Table 3.
Mean, Range and Median of haemoglobin of the participants enrolled in the study.

Discussion: Hb was found to be ranged from 6.0 to 22.80 g/dl, mean ± SD of Hb being 12.63 ± 2.91 g/dl.

11. Advances in colorimetry

The accessibility of smartphones and the speedy development of different mobile applications is having a great effect in our daily life [13]. Nevertheless, till now, smartphones have been rarely applied to other sciences and their applications in this research field may provide an immense leap in terms of routine sample analysis with an accessible, fast, simple and low-cost strategy. Smartphone cameras produces images using a red, green, and blue (RGB) color code. Based on this code, simple analysis of these digital pictures to quantify color changes is done with the help of different applications. These differences in color have been used for the quantification of various analytes in liquid suspensions, both in simple solutions and direct sampling matrices [14]. In many cases, a reagent is used to produce a colored compound through the reaction with a particular analyte, and the color of the newly formed analyte-reagent complex is measured [15].

Smartphones are embedded with an optical chemical or biosensing base via the use of an attached lighting control enclosure that is fitted to the device and the assay platform that may be strip, cassette, cuvette etc. In this configuration, the smartphone can function as an illumination source (Light Emitting Diode), a signal detector, and a signal processor. Based on the conditions of the assay, various lenses, filters, diffraction gratings, and alternative light or power sources may be incorporated within the attached enclosure to intensify signal detection from colorimetric, fluorescence-based, chemi/bioluminescence-based, and scattering-based assays [16, 17].

Pharmaceutical compounds can be determined in dosage forms and biological matrices by using various colorimetric reagents with the help of spectrophotometric and chromatographic methods. The colorimetric methods have more sensitivity than that of ultraviolet spectroscopic methods. Moreover colorimetric methods measure light of longer wavelengths which decreases effects of interfering excipients. Most of the procedures are inexpensive and reagents easily available. There is a great scope for development of new reagents and new colorimetric methods [18].

There has been a focus on newly developed chemosensors on systems that permit for portable, on-site testing of the target analytes without requiring high yield, costly laboratory instrumentation [19, 20]. It is challenging to design a portable chemosensors to maintain high selectivity, sensitivity, and wide applicability in the efficient measurement of various analytes especially that are similar in structure and size. The Levine group has developed sensitive and selective fluorescence-based systems for analyte detection by using the property of cyclodextrin to act as a supramolecular scaffold that aids proximity-induced, highly analyte-specific interactions between an analyte of interest and a high-quantum yield fluorophore [21–25].

Recently there has been developments in the the field of optofluidics making use of light matter interactions in integrated devices with significantly improved features [26]. The newly developed microfluidic device integrated in optofluidic system has been introduced. The sensitivity of measuring concentration of sample by optofluidic system is better than that of other methods [27].

Author details

Yagya Kumari Shrestha[1]* and Shree Krishna Shrestha[2]

1 KIST Medical College and Teaching Hospital (KISTMCTH), Nepal

2 National Trauma Center, National Academy of Medical Science (NAMS), Nepal

*Address all correspondence to: dr.yagyashrestha@gmail.com

References

[1] Guide to colorimeter. Available from: https://www.sherwood.scientific.com

[2] Outa N, Robertson AR. Colorimetry: Fundamentals & Applications. 2nd ed. Great Britain: IS & T; 2005, Chapter 1, Light, Vision, & Photometry; 1-36. 2

[3] Chatterjee MN, Shinde R. Textbook of Medical Biochemistry. 7th ed. New Delhi: Jaypee Brother Medical Publishers Private Limited; 2007

[4] The National Bureau of Standards transferred this example to the Smithsonian in 1965. A tag reads "BAUSCH & LOMB OPTICAL CO./ ROCHESTER, U.S.A."

[5] Warner DJ. The Duboscq colorimeter. Bulletin of the Scientific Instrument Society. 2006;**88**:68-70

[6] Stock JT. The Duboscq Colorimeter and its Inventor. Journal of Chemical Education. 1994;**71**:967-970

[7] Kricka LJ, Park JY. Optical techniques. In: Burtis CA, Ashwood ER, Bruns DE, editors. Tietz Textbook of Clinical Chemistry and Molecular Diagnostics. 5th ed. Philadelphia: Saunders; 2012. pp. 242-244

[8] Wilson K, Walker J, editors. Principles and Techniques of Biochemistry and Molecular Biology. 7th ed. New York: Cambridge University Press; 2010. pp. 477-519

[9] Upadhyay A, UpadhyayK NN. Biophysical Chemisty: Principles and Techniques. 4th ed. Gurgaon: Mumbai; 2014

[10] Upadhyay A, UpadhyayK, Nath N. Techniques in Biochemistry (1st ed). Keshav Puram, Delhi: Ahuja Publishing House. 2010

[11] Bishop ML, Fody EP, Schoef LE. In: Bishop CML, Fody EP, Schoeff LE, editors. Clinical Chemistry Techniques, Principles, Correlations. 6th ed. 2010. pp. 130-140

[12] Shrestha YK, Khanal MP, Shrestha SK, Shrestha J, Shrestha N, et al. Blood Lead level among children between 8-18 years of age by atomic absorption spectrophotometry: A descriptive cross-sectional Study. JNMA; Journal of the Nepal Medical Association. 2021;**59**(242):1012-1016

[13] Afonso YF, Salas G, Barahona IF, Herranz F, Gruttner C, Fuente JMDL, et al. Smartphone-based colorimetric method to quantify iron concentration and to determine the nanoparticle size from suspensions of magnetic nanoparticles. Wiley-Vch. 2020;**37**(7): 1-22

[14] Levin S, Krishnan S, Rajkumar S, Halery N, Balkunde P. Science of the Total Environment. 2016; **551–552**:101

[15] Oncescu V, Mancuso M, Erickson D. Lab on a Chip. 2014;**14**:759

[16] McCracken KE, Yoon JY. Analytical Methods. 2016;**8**:6591

[17] Cornell RM, Schwertmann U. The iron oxides: Structure, properties, reactions, occurrences, and uses, Wiley-VCH, 2003 noparticle size from suspensions of magnetic nanoparticles. Wiley-Vch. 2020;**37**(7):1-22

[18] Gummadi S, Kommoju M, et al. Colorimetric approaches to drug analysis

and applications – A review. American Journal of PharmTech Research. 2019; **9**(01):14-37

[19] Haynes A, Halpert P, Levine M, et al. Colorimetric detection of aliphatic alcohols in β-cyclodextrin solutions. ACS Omega. 2019;**4**(19):18361-18369

[20] Huang X, Xu D, Chen J, Liu J, Li Y, Song J, et al. Smartphone-based analytical biosensors. Analyst. 2018;**143**: 5339-5351

[21] DiScenza DJ, Culton E, Verderame M, Lynch J, Serio N, Levine M. Towards rational Chemosensor design through improved understanding of experimental parameter variation and tolerance in cyclodextrin-promoted fluorescence detection. Chem. 2017;**5**:34–34/15

[22] Serio N, Miller K, Levine M. Efficient detection of polycyclic aromatic hydrocarbons and polychlorinated biphenyls via three component energy transfer. Chemical Communications. 2013;**49**:4821-4823

[23] Serio N, Moyano DF, Rotello VM, Levine M. Array-based detection of persistent organic pollutants via cyclodextrin promoted energy transfer. Chemical Communications. 2015;**51**: 11615-11618

[24] DiScenza DJ, Levine M. Sensitive and selective detection of alcohols via fluorescence modulation. Supramolecular Chemistry. 2016;**28**: 881-891

[25] Li Z, Askim JR, Suslick KS. The optoelectronic nose: Colorimetric and fluorometric sensor arrays. Chemical Reviews. 2019;**119**:231-292

[26] Chen C, Mehl BT, Munshi AS, Townsend AD, Spence DM, Martin RS. 3D-printed microfluidic devices: Fabrication, advantages and limitations —A mini review. Analytical Methods. 2016;**8**(31):1-17. DOI: 10.1039/ c6ay01671e

[27] Dali MRM et al. Concentration measurement of coloured solution using optofluidic approach. Journal of Physics: Conference Series. 2022;**2411**:012005

Chapter 3

Application of UV-Visible Spectrophotometric Colour Analysis in Different Natural Product Identification

Sonia Karuppaiah, Nithya Sermugapandian, Sumithra Mohan and Manikandan Krishna

Abstract

When light falls on a substance, a part of the light that interacts with a substance is absorbed, and the remaining light is either reflected or transmitted through the substance. Colorimetric is used for the detection of dyes, food colourant in which the wavelength of light that different coloured dyes absorb varies. The majority of dyes are conjugated substances with alternate double and single bonds, and they generally absorb light in the visible spectrum. It is used for the food colourant studies due to the long history and widespread use of dyes in retail goods, the spectroscopic identification of food colourings appears to be a viable methodology.

Keywords: colorimetric, dyes, food colourant, natural dyes, reflected

1. Introduction

When light falls on a substance, a portion of the light that interacts with a substance is absorbed, and the remaining light is either reflected or transmitted through the substance. Visible light is reflected by objects that we perceive as having colour. The wavelength of light that is reflected determines the colour of the material that we can perceive [1]. A material that seems blue to us will reflect light in the visual spectrum's blue region (430–480 nm). The same substance absorbs light that is complementary to the light that is reflected, according to the colour wheel. As a result, the blue material absorbs light in the visible spectrum's orange band (590–630 nm). Because not all chemicals absorb in the visible area, they appear colourless to the human eye.

Colour measurements are a way to quantify the values of the hues that humans perceive. Measurements of colour are affected by lighting, an object's spectral properties, and the human eye's spectral sensitivity properties. A colour value can be estimated if the spectrum reflection of the item is known since the spectral distribution of the illumination and the spectral sensitivity characteristics (colour-matching

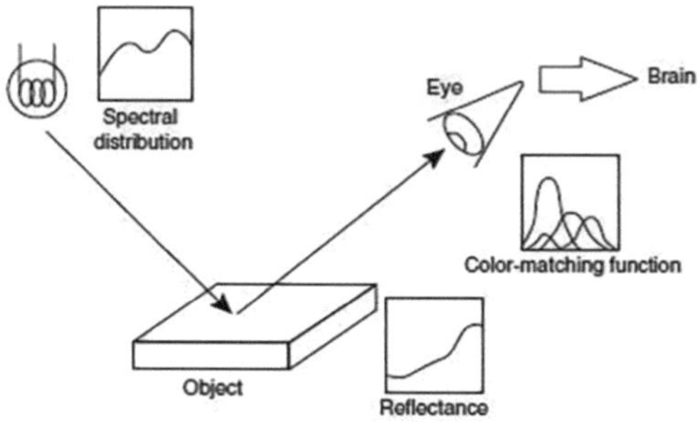

Figure 1.
What is colour measurement?

function) of the eye are established in the JIS standards. (The computation can use spectral transmittance if the light can flow through the item. However, the explanations that follow employ spectral reflectance.) To further clarify, the JIS standard calculates the spectral distribution of the illumination and the colour-matching function under a number of different situations in **Figure 1**.

2. Light

Light is defined by its energy, E, and its wavelength, λ. Here, c is the speed of light and h is Planck's constant [2].

$$E = \frac{hc}{\lambda} \tag{1}$$

The relationship between light's energy and wavelength is inverse. Therefore, light with increased energy has a shorter wavelength.

3. Spectrophotometer

A UV–Visible (UV–Vis) spectrophotometer is used to experimentally test light absorption. This device makes use of a light source a particular monochromator turns into a range of wavelengths that control over a sample and towards a detector on the spare end. If the organic chemical is a solid, a solvent is essential since the samples must be in a liquid. The container warned to carry this solution is a cuvette. The cuvette may be formed of quartz crystal, glass, or plastic and have a certain path length depending on the sample. The distance that light must travel over the sample is measured in terms of path length. A sample blank of only the solvent is needed because it will also absorb light. In order to display just the absorbance induced by the sample, the instrument can remove the background spectrum of the solvent from the absorbance spectrum of the sample component. The percentage of the primitive light that passes through the sample is known as the transmittance, or T. The irradiance, or efficiency per second per unit area, of the light beam, just before it hit the sample is P0 in this case. P denotes the intensity of the light beam that hits the detector [3]. Due to the sample's absorption of some light, P is normally lower than P0.

Absorbance is a negative log of transmittance.

$$A = \log\left(\frac{P_0}{P}\right) = -\log T \quad (2)$$

The value of absorbance ranges from 0 (no absorption) to 2 (99% absorption). When no light is absorbed, P0 equals P and transmittance equals one. As a result, absorption is zero. If 90% of the light is absorbed, 10% is transmitted, and T = 0.1. This yields an absorbance of one. If 99% of the light is absorbed, then 1% of the light is transmitted (T = 0.01), and absorbance equals 2.

The resulting spectrum is a plot of absorbance versus wavelength. This range for a UV–Vis spectrophotometer is between 200 and 800 nm.

4. Beer–Lambert law

The transmittance and absorbance of a certain substance are proportional to its concentration in solution. The Beer–Lambert law describes this relationship.

$$A = \varepsilon cl \quad (3)$$

The sample absorbance is equal to the product of the chemical concentration, route length, and molar attenuation coefficient. This coefficient is specific to each molecule and varies with wavelength. However, if the wavelength is maintained fixed, the molar attenuation coefficient remains constant regardless of concentration changes [3]. The wavelength with the maximum absorbance of the sample, known as max, will likewise have the highest molar attenuation coefficient (**Table 1**).

5. Dyes

The wavelength of light that different coloured dyes absorb varies. The majority of dyes are conjugated substances with alternate double and single bonds, and they generally absorb light in the visible spectrum.

The conjugated portion of the dye molecule might be lengthy, indicating a high degree of conjugation with numerous alternating double and single bonds, or extremely short, indicating a low degree of conjugation with few alternating double and single bonds. It's not necessary for these alternating double bonds to merely be between two carbons. The carbonyl groups and the double bonds between carbon and oxygen can be examples of these conjugated bonds. The amount of conjugation affects how much light the chemical can absorb at each wavelength. For instance, substances that have higher levels of conjugation absorb light at longer wavelengths than substances that have lower levels of conjugation.

	Absorbance range (nm)	**λ_{max} (nm)**
Fluorescein		
β-carotene		
Indigo		

Table 1.
Absorbance and λ_{max} of dyes.

Delocalized electrons inhabit molecular orbitals according to molecular orbital theory. The highest energy orbital with an electron is known as the highest occupied molecular orbital or HOMO. The lowest energy orbital without an electron is known as a LUMO or the slightest unoccupied molecular orbital. The energy difference between the HOMO and LUMO is often quite big in molecules with little to no conjugation [4]. However, the efficiency difference between the HOMO and LUMO is lower for conjugated molecules.

The molecule must absorb light with energy equivalent to the initiative gap between the two orbitals in order to excite a transition of an electron from one energy level to another, or from the HOMO to the LUMO. For this reason, higher energy light —such as UV light—is necessary to excite an electron in molecules with a wide energy gap. However, because dyes have a narrower energy gap, an electron must be excited by lower-energy light, like visible light.

$$\Delta E_{(HOMO-LUMO)} = \frac{hc}{\lambda} \tag{4}$$

For this reason, higher energy light—such as UV light—is necessary to excite an electron in molecules with a wide energy gap. However, because dyes have a narrower energy gap, an electron must be excited by lower-energy light, like visible light. Remember that the relationship between light's energy and wavelength is inverse [5]. Therefore, light with more energy has shorter wavelengths than light with lower energy, which has longer wavelengths.

5.1 Natural dyes

The term "natural dyes" refers to colourants (including dyes and pigments) that are derived from plant or animal matter without the use of chemicals. Although there are some known vat, solvent, pigment, and acid types, they are primarily mordant dyes. Natural dyes are used to colour clothing, food, medicine, and cosmetics. Additionally, very little amounts of dyes are used to colour paper, leather, shoe polish, wood, cane, candles, etc.

In the past, only natural sources were used to create dyes. But because dyes obtained from comparable plants or other natural sources are affected and subjected to the whims of climate, soil, cultivation techniques, etc., natural dyes suffer from some inherent drawbacks of standardised application and the standardisation of the dye itself. In order for natural dyes to be genuinely commercialised and to compete favourably with synthetic dyes, standardisation techniques are crucial and play a huge role. Indian culture has a long history of employing natural dyes. Traditional skilled craftspeople in the nation have been manufacturing natural dyed textiles for centuries in numerous villages. When used alone, natural dyes have several restrictions on their colour fastness and brilliance. However, they provide vibrant and quick hues when used with metallic mordants. Although using metallic mordants is not always environmentally benign, the pollution issues they cause are of very low order and are easily resolved. Therefore, "Mild Chemistry" can be used to obtain practically identical results in place of environmentally harmful technology for the production of colours.

This chapter discusses the self-association of certain dyes and divides the phenomenon into two categories: H- and J aggregates. The aggregation processes of the following dyes, which have been the subject of spectroscopic studies: rhodamine B, rhodamine 6G, Neutral Red, Nile Blue A, Safranine T, Thionine, Methylene blue, Methylene green, thiazole orange, and TO-3, were discussed. One of the characteristics of dyes in solution is their capacity for self-association, which many colours exhibit. Aggregation is a phenomena that changes the characteristics of dyes as well as their coloristic and photophysical properties. Due to high intermolecular attractive forces between the molecules, the aggregation phenomena in solution or at the solid–liquid interface is a frequently encountered event in dye chemistry. The ionic dyes have a well-known propensity to congregate in diluted solutions, resulting in the formation of dimers and occasionally even higher order aggregates. In this scenario, factors including dye concentration, dye structure, ionic strengths, temperature, and the presence of organic solvents all have a significant impact on the molecular nature of the dye. Although the structure and behaviour of dyes are highly idiosyncratic, there are some fundamental guidelines for aggregation that have been established. It may rise with an increase in dye concentration or ionic strength; fall with an increase in temperature or the addition of organic solvents; fall with the addition of ionic solubilising groups to the dye structure; fall with the addition of long alkyl chains; fall with an increase in aggregation due to higher hydrophobic interaction in solution.

In both fundamental science and technological applications including optical memory, organic solar cells, and organic light emitting diodes, dye aggregates have played a significant role [6–8]. Among all synthetic dyes, xanthene dyes are among the earliest and most widely used. They were used for food colouring and clothing, among other things [9]. The unique photophysics characteristics of these kinds of molecules are the reason for their numerous and expanding applications in physics and chemistry. Because of their high time-zero anisotropy, photostability, and red emission, they are used as fluorescent protein probes in detecting protein orientation in biology and as probes in biochemistry to monitor membrane fusion, determine the distance between aggregations, and detect protein orientation.

It was discovered in the 1930s that a group of dyes known as cyanines have a high propensity to aggregate in polar solvents, and this group has since been the focus of numerous studies that primarily analyse the effects of concentration on absorption and emission spectra. The dimerization process, which takes place in a specific concentration range and occurs before the development of more complex aggregates, was given the most attention by the bulk of the authors. Spectral sensitization in photography, size-enhanced nonlinear optical polarizabilities, sensitization of semiconductor materials, etc. are just a few technologically interesting processes that include organised assemblies of cyanine dyes as molecular functional units. In these situations, aggregates develop at the air-water interface on solid surfaces or in monomolecular layers, where the packing of the chromophores is aided by particular dye-substrate interactions.

6. Measurement of colour

Measure the object's spectral reflectance before using a UV–VIS spectrophotometer to measure colour. The colour is expressed as a numeric number using calculations based on the spectrum distribution of the illumination, the object's measured spectral reflectance, and the colour-matching function. To acquire colour measurement values when the spectral reflectance spectrum is measured, the colour measurement programme stores illumination spectral distributions and colour-matching function values [10]. The XYZ tristimulus values serve as the foundation for measuring colour. "Methods of Colour Measurement – Reflecting and Transmitting Objects" determines the XYZ tristimulus values.

7. Colour difference

A colour specification system is a way to define colours as numerical values, whereas a colour difference is a way to express how different hues are from one another [11]. The Uniform Colour Space (UCS), which is more closely aligned with the human visual system, is used in calculations to describe colour difference values mathematically. A common UCS colour specification system is the L*a*b* colour space. A* and b* stand for hue and saturation, whereas L* stands for brightness. The technique of computation in the L*a*b* colour system is shown in JIS Z 8729, "Colour specification – CIELAB and CIELUV colour spaces." The L*a*b* value for each colour of the object (sample) is used to determine the colour difference. Using "Colour specification – Colour differences of object colours," the colour difference E*ab in the L*a*b* colour system is calculated.

8. Absorbance wavelength comparison

To scrutinise the absorption properties of fluorescein, −carotene, and indigo dye using UV–Visible absorption spectroscopy, besides known as UV–Vis spectroscopy. A compound between a light source and a photodetector in order to perform UV–Vis spectroscopy. When light's wavelengths match the energies required to excite a molecule's electrons, it will absorb that light and scatter or transmit the rest.

- The fluctuation in the number of photons of each wavelength that reach the detector is represented by an absorption spectrum. There are fewer photons of that wavelength detected when the absorbance is higher. A spectrum of the true solvent, whichever the device determines then subtracts from the spectrum of the dye solving to only display the dye's absorbance information.

- The colours in a fume hood use -carotene and indigo solutions like hexane and dimethylformamide, respectively. To get solvent on thin nitrile gloves, such as DMF or hexane, change them immediately [12].

- Typically, a cuvette will have two see-through sides and two artificial, opaque sides. The cuvette's lucent sides can easily get dirty, thus it's best to always hold it by the rough sides.

- Examine a fluorescein-water mixture for flaws. About three-quarters of your cuvette should be filled with deionised water to create your solvent blank.

- Use a laboratory wipe to clean the cuvette's see-through sides. If you notice any air bubbles in your cuvette, you can get rid of them by giving it a light tap.

- Finally, open the spectrometer's sample cavity and remove the fragile cuvette. Dust and fingerprints can skew results by either absorbing or reflecting light.
- Place the cuvette into the representative holder with the lucent side facing the light. Put a heavy lock on the isolation room door. Scan your cuvette as the solvent blank using the spectrometer's software as directed. The absorbance range should be from 200 to 800 nm.
- When the scan is finished, you should remove any large drops of water by emptying the cuvette and the associated compressed air.
- Then, take it to the common hood and fill the cuvette three-quarters of the way with the 3 mM fluorescein solution using a clean Pasteur pipette.
- Before bringing the cuvette to the spectrometer, make sure there are no air bubbles in it and pop them if there are. Before inserting the cuvette into the sample chamber, make sure the see-through sides are spotless. The sample cuvette should be scanned.
- When the scan is complete, figure out which wavelengths fluorescein was able to absorb. Maximum absorbance occurs at a specific wavelength, denoted by "max."
- Write down the absorbance peaks' wavelengths and the full absorption spectrum, then store this information.
- Then, grab a clean cuvette and fill it three-quarters of the way with hexane to use as a blank solvent.
- The cuvette should be placed in the spectrometer after any air bubbles have been removed and the transparent sides have been cleaned. Perform a solvent blank scan using the cuvette and an absorbance scanner set to 200–800 nm.
- Remove any remaining fluorescein from your cuvette by emptying it into the aqueous trash and rinsing it thoroughly with water.
- Next, you'll need a hexane solvent blank to analyse beta-carotene. After removing any trace of water from the cuvette, fill it three quarters of the way with pure hexane.
- Remove any trapped air, wipe off. Place the cuvette in the light path of the spectrometer. As if the cuvette were a solvent blank, perform a scan with the absorbance set at 200–800 nm.
- When the scan is finished, discard the cuvette and blow it with compressed air to eliminate any residual microorganisms.
- Fill the cuvette 3/4 of the way with the 3 mM -carotene solution and return it to the spectrometer using a good pipette.

- Remove the cuvette's lucent sides, insert it into the spectrometer, and perform a scan. Notate in your lab book how long in wavelength each peak was.

- Now, let us have a look at the indigo dye. Since the indigo solution is in DMF, you'll need to dispose of the used cuvette in the organics bin and flush any residual beta-carotene or hexane from it with pure DMF [13]. Then, you'll fill the cuvette up to three-quarters of its capacity with pure DMF, replace it in the spectrophotometer, and make sure that your scan settings have not changed. Once the solvent blank scan is complete, you can toss the cuvette's contents into the compost bin and blow dry the remaining moisture.

- Pipette 3/4 full of the given 3 mM indigo dye solution into the cuvette, and then scan it as your sample. Document your absorbance results and store them for further use.

- Finally, after rinsing it out with acetone, discard the cuvette's contents into the compost or garbage.

8.1 Measurement using UV visible spectroscopy

- Fluorescein, beta-carotene, and indigo dye are studied using absorption spectroscopy, also known as UV–Vis absorption spectroscopy. In UV–Vis spectroscopy, a solution is placed between a light source and a photodetector. The wavelengths of light required to activate their electrons will be absorbed, and the remaining light will be diffused or transmitted [14].

- An absorption spectrum depicts the fluctuation in the number of photons of each wavelength that reach the detector.

- When the absorbance is high, fewer photons of that wavelength are picked up by the detector. To get the dye absorbance information, you need a spectrum of the pure solvent, which the instrument will subtract from the spectrum of the dye solution. An example of what is called a solvent blank [15]. The -carotene and indigo solutions are prepared in a fume hood using hexane and dimethylformamide, respectively.

- Typically, cuvettes have two clear sides and two rough or opaque sides. To keep the clear sides clean, always hold the cuvette by the textured sides.

- Examine a fluorescein solution in water. Fill your cuvette 3/4 full with deionised water to create your solvent blank.

- Clean the clear sides of the cuvette using a lab wipe. Inspect the cuvette for air bubbles and tap it slightly to get rid of them.

- After that, take the cuvette to the spectrometer. Open the spectrometer's sample chamber, then wash your cuvette once again. Smudges and dust can reflect or absorb light, which might result in inaccurate statistics. Now, position the cuvette in the sample holder so that its transparent sides are facing the light source. Completely shut the sample room door. Set up a scan of absorbance between 200 and 800 nm in your spectrometer's software, using your cuvette as the solvent blank.

- After the scan is finished, drain the cuvette and use compressed air to blast large water drops away.
- After that, take it to the common hood and add the provided 3 M to the cuvette until it is 3/4 filled using Pasteur pipette.
- Before taking the cuvette to the spectrometer, detect air bubbles and get rid of them if necessary. Recall to before placing the cuvette into the sample chamber, clean the cuvette's transparent sides one last time. The next step is to scan the cuvette.
- After the scan is finished, note the wavelengths that fluorescein absorbed. The frequency of the wavelength with the greatest absorption is referred to as max.
- In your lab notebook record and save the data after noting the wavelengths of the absorbance peaks and the total absorption range.
- Empty the cuvette into the watery trash and rinse it to eliminate any remaining fluorescein.
- To analyse - carotene next, you'll require a blank for hexane solvent. Before adding pure hexane to the cuvette, rinse it several times with hexane to get rid of any water that may still be present.
- Clear the cuvette of any air bubbles and place it inside the spectrometer. Ensure that the scan is adjusted to absorbance between 200 and 800 nm. The cuvette will then be scanned as a solvent blank at nm and then scan the cuvette as a solvent blank.
- After the scan is completed, Dump the cuvette into the organic waste after the scan is finished, then pressurise the air to dry it.
- Using a clean pipette, fill the cuvette 3/4 of the way with the supplied 3 M - carotene solution before returning to the spectrometer.
- Clean the clear sides of the cuvette, place it in the spectrometer, and use it as your sample to be scanned. Write the wavelengths of the peak in your lab notebook and save the data.
- Examine the indigo dye right now. Because the indigo solution is in DMF, place your cuvette in the organic trash. Then, thoroughly rinse away any remaining hexane and -carotene with pure DMF. Fill the cuvette 3/4 full with pure DMF, place it in the spectrometer, and repeat the previous scan conditions. After the solvent blank scan, immerse the cuvette in organic waste and dry it with compressed air. The provided 3 M indigo dye solution into the cuvette until it is 3/4 filled, then scan it as your sample.

- Finally, rinse your cuvette with acetone after emptying it into the organic garbage.

8.2 Applications

8.2.1 Applications of UV spectroscopy: Identification of dyes

A “colour wheel” is created by arranging several hues in a circle. One potential solution to this is depicted in the diagram. Complementary colours are those that are settled opposite one another on the colour wheel [16]. Colours that complement one other include blue and yellow, red and cyan, green and magenta, and red and indigo. White light is created by combining two light hues that are complementary to one another in **Figure 2**.

8.2.2 Natural dyes extraction

Anthocyanin is a natural pigment found in purple cabbage (Brassica oleracea var. capitata f. rubra). This dye is made up of hydroxyl and carbonyl groups. Spinach (Spinaciaoleracea) includes natural chlorophyll pigments called Chlorophyll a and Chlorophyll b. Anthocyanin and chlorophyll pigments are extracted from purple cabbage and spinach by cutting them into small pieces and mashing them into a paste with a mortar [17]. The samples are then placed separately in an ultrasonic cleaner for 15 minutes at a frequency of 37 Hz in the ‘degas’ mode at 30°C. The coloration of the samples is separated by centrifugal force at 2500 rpm for 30 minutes. The pH of the produced anthocyanin and chlorophyll dyes in ethanolic solution is 7.03 and 7.2, respectively [18]. Now, in a 1:1 ratio, combine the extraction of purple cabbage juice and spinach juice, and then add Ethanol to the mixture. After 10 minutes, sieve the mixture. The pH of the produced dye is 7.15.

8.2.3 Spectral studies of food colourants

Due to the long history and widespread use of dyes in retail goods, the spectroscopic identification of food colorings appears to be a viable methodology. The spectrum information for frequently used dyes that is necessary as a result is presented for

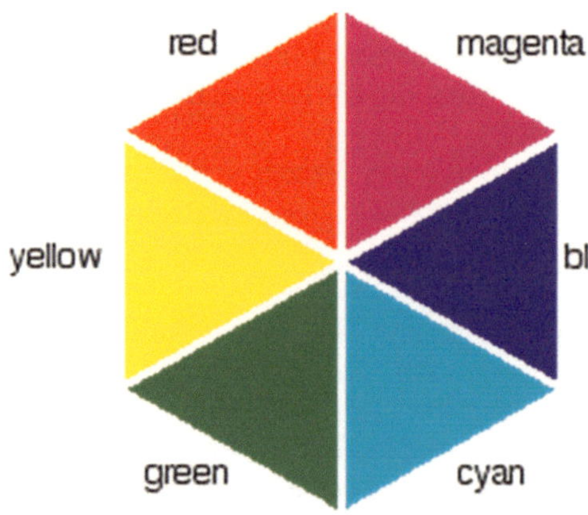

Figure 2.
Complementary colours in colour wheel.

instructional purposes. The analysis of food colorings in four distinct lemonades and chocolate beans, both qualitatively and quantitatively, has been used to educate students to key analytical methods such sample preparation and the elimination of confounding variables. In the cases of tartrazine and curcumin, these analyses also show the method's limitations in the visible light spectrum. Quantitative studies can study typical concentration calculations by using Lambert–Beer–Bouguer's Law in various variations.

Most people begin eating with their eyes. Consumers today expect food to look as appetising as possible, which can be achieved by adding extra or artificial colouring. Plant extracts or saffron were utilised as natural dyes [17]. Prior to industrialisation, this form of natural food colouring was out of reach for the average person. Under favourable economic circumstances, it was necessary to recolor food with dyes due to the somewhat unpleasant colour shift that occurred during the processing of food.

Because of their affordable production costs, greater colouring capabilities, and practical application in products, these new artificial food dyes have replaced both dangerous inorganic salts and natural colourants. Self-service supermarkets, mass production, and transparent packaging have all increased, making it necessary to standardise the Food colour schemes were established. Contrary to what people might think, many of these are made synthetically. We investigated carotenes, chlorophylls, and curcumin in diet in this work.

It is necessary to determine attenuation coefficients for systems with practical relevance - like lemonade or candies - when teaching quantitative analysis by spectroscopy. For support, extensive experimental data pertaining to attenuation coefficients of food dyes are provided as supplementary materials. Below are the results of spectroscopic analysis of dyes such as Allura red, Amaranth, Apocarotenal, Azorubine, Beta carotene, Brilliant blue, Caspasnthin, Curcumin, Carmine, Erythrosine, Green S, Indigo carmine, and Lutein.

8.2.4 Calibration of dyes - β-carotene calibration curve

UV–Vis spectroscopy was used by comparing the absorbances of five different concentrations of -carotene solutions, it was possible to link absorbance intensity and concentration for this compound. Make an absorbance vs. -carotene concentration calibration curve, and then determine the equation that describes the connection [18].

- Hexane should be used as your solvent blank in a clean cuvette that is 3/4 full.

- Clear the cuvette's clear sides of any air bubbles, then place it in the spectrometer. Set up an absorbance scan with the cuvette serving as the solvent blank between 200 and 800 nm.

- Dump the cuvette into the organic waste after the scan is finished, then pressurise the air to dry it.

- Pour the 1.9 M -carotene solution, which has the lowest concentration, into the cuvette using a clean pipette.

- Fill the cuvette 3/4 full with the 1.9 M -carotene solution, which has the lowest concentration, using a clean pipette.

- Before placing the cuvette into the spectrometer and scanning it as a sample, clean the sides of the cuvette.
- Display the absorbance strength of the peak at 450 nm after the scan is finished. In your lab notebook, write the absorbance for the 1.9 M solution.
- Submerge the cuvette in the organic waste, followed by a hexane rinse and air drying.
- Using a clean pipette, add the 3.7 M -carotene solution to the cuvette until it is 3/4 full. Before placing the cuvette into the spectrometer and scanning it as another sample, clean its sides.
- After the scan is finished, the peak's absorbance at 450 nm is calculated. Rep for the remaining choices in order of decreasing concentration. Save your work after you are done.
- Finally, immerse the cuvette in organic waste and clean it with acetone.

The second half uses the absorbances of five distinct solutions containing various concentrations of -carotene, UV–Vis spectroscopy was used to establish a relationship between absorbance intensity and concentration. The equation expressing that relationship is derived from a calibration curve of absorbance versus -carotene concentration in **Table 2**.

- Create a table with columns for -carotene concentration and absorbance at 450 nm in your lab notebook to start.
- After the scan is complete, empty the cuvette into the organic garbage and blow it dry.
- Use a clean pipette to dispense the 1.9 M -carotene solution, the lowest concentration, into the cuvette to a depth of 3/4 filled.

Trim the cuvette's sides, place it in the spectrometer, and scan the sample there.

- Display the peak at 450 nm's absorbance intensity after the scan is complete. This should be noted as the absorbance for the 1.9 M solution in your lab notebook.

	Concentration (μM)	**Absorbance at λ_{max} (450 nm)**
1	1.9	
2	3.7	
3	7.5	
4	11	
5	15	

Table 2.
Absorbance of β-carotene.

- Next, pour the contents of the cuvette into the organic waste, wash it with hexane, and then seal it with air pressure.
- Use a clean pipette to dispense the 3.7 M -carotene solution into the cuvette until it is 3/4 full. Clean the cuvette's sides before placing it in the spectrometer and scanning it like a different sample.
- After the scan is complete, note the peak's 450 nm absorbance. Working from lowest to greatest concentration, repeat this procedure for the remaining solutions. When finished, save your data.
- Afterward, rinse the cuvette with acetone after emptying it into the organic garbage.

8.2.4.1 Results

Blue and purple light is also absorbed by -carotene. In hexane, beta-carotene has a maximum absorbance of 450 nm and another significant peak at 478 nm. -carotene appears orange in part because of the intense absorption of purple light.

DMF's indigo absorbs red, orange, yellow, and UV light, with a clear peak at 611 nm. In order to get its distinctive colour, indigo dye reflects mainly blue and purple light.

The concentration calibration curve for -carotene. Plot the congregations of the liquids versus the -carotene absorbance values at 450 nm from the second part of the experiment. Execute a trend line next, and then locate the linear equation that best describes the data. In this equation, y represents absorbance, x represents concentration, and the slope is uniform due to the conforming molar attenuation coefficient and the slope factor. Path length, in accordance with the Beer–Lambert law. Fill in the -carotene absorbance at 450 nm from the first fraction of the lab and rearrange it to solve for concentration. Nearly 3 M/L of concentration must have been computed. Complementary Colours.

8.2.4.2 Interpretation for dyes

The wavelength at which fluorescein absorbs blue and purple light in water is 490 nm. Red light is not absorbed by it; it only takes a little amount of yellow and green light. Fluorescein solids are red, while fluorescein liquids are often yellow to green. Blue and purple light are also absorbed by -carotene [19]. In hexane, the greatest absorbance of -carotene is 450 nm, with another big peak at 478 nm. Because of due to a high amount of purple light absorption, −carotene appears orange. With a distinct peak at 611 nm, indigo absorbs UV light as well as red, orange, and yellow light. Dye indigo as a result, reflects mostly blue and purple light, giving it its distinctive colour.

- For the -carotene concentration calibration curve, Plot the -carotene absorbance values at 450 nm vs. the solution concentrations [19].
- Next, use a trend line to identify the linear equation that best matches the data. In this equation, y is absorbance, x is concentration, and the slope is the product of the applicable molar attenuation coefficient and the pathlength, according to the Beer Lambert law., according to the Beer–Lambert law.

8.2.5 Spectroscopic analysis of methylene blue

In the textile industry, Methylene Blue (MB) is widely used to dye wool, cotton, and silk [6]. It can also be used to dye specific body tissues and fluids before or during surgery. The visible light absorption spectra analysis confirmed that methylene blue has high order methylene blue aggregate without any preconception about the aggregation order.

9. Conclusion

Since Colorimetric is widely used not only in the pharmaceuticals but also used as food colourant, dyes and natural dyes estimation.

Author details

Sonia Karuppaiah[1]*, Nithya Sermugapandian[2], Sumithra Mohan[3] and Manikandan Krishna[4]

1 Department of Pharmaceutical Chemistry, Sri Ramachandra Faculty of Pharmacy, SRIHER (DU), Chennai, India

2 Department of Pharmacology, Sri Ramachandra Faculty of Pharmacy, SRIHER (DU), Chennai, India

3 Department of Pharmacology, SRM College of Pharmacy, SRMIST, Kattankulathur, India

4 Department of Pharmaceutical Analysis, SRM College of Pharmacy, SRMIST, Kattankulathur, India

*Address all correspondence to: soniapharm68@gmail.com

References

[1] Was-Gubala J, Starczak R. Nondestructive identification of dye mixtures in polyester and cotton fibers using Raman spectroscopy and ultraviolet-visible (UV-vis) micro spectrophotometry. Applied Spectroscopy. 2015;**69**(2):296-303

[2] Westland S, Laycock K, Cheung V, Henry P, Mahyar F. Colour harmony. Colour: Design & Creativity. 2007; **1**(1):1-5

[3] Demissie EG, Woyessa GW, Abebe A. UV/Vis spectrometer determination of caffeine in green coffee beans from Hararghe, Ethiopia, using beer-lambert's law and integrated absorption coefficient techniques. Scientific Study & Research. Chemistry & Chemical Engineering, Biotechnology, Food Industry. 2016; **17**(2):109

[4] Khan SU, Mahmood A, Rana UA, Haider S. Utilization of electron-deficient thiadiazole derivatives as π-spacer for the red shifting of absorption maxima of diarylamine-fluorene based dyes. Theoretical Chemistry Accounts. 2015;**134**:1-7

[5] Richhariya G, Kumar A, Tekasakul P, Gupta B. Natural dyes for dye sensitized solar cell: A review. Renewable and Sustainable Energy Reviews. 2017;**69**: 705-718

[6] Goftar M, K, Moradi K, Moradi N. Kor2: Spectroscopic studies on aggregation phenomena of dyes. European Journal of Experimental Biology. 2014;**4**(2):72-81

[7] Gräb P, Geidel E. Spectroscopic studies of food Colorings. World Journal of Chemical Education. 2019; 7(2):136-144. DOI: 10.12691/wjce-7-2-13

[8] Gräb P. Spectral Data of Food Colorings. 2019 [Online]. Available from: https://www.chemie.uni-wuerzburg.de/didaktik/lehrer/experimente-zurspektroskopie/

[9] Fernandez-Pérez A, Marba G. Visible light spectroscopic analysis of methylene blue in water; what comes after dimer? ACS Omega. 2020;**5**:29801-29815

[10] De León-Rodríguez LM, López-de-Alba PL, García-Campos R, De León-Rodríguez LM. Simultaneous determination of methylxanthines in coffees and teas by UV-vis spectrophotometry and partial least squares. Analytica Chimica Acta. 2003; **493**(1):83-94

[11] Riu J, Schonsee I, Barcelo D. Determination of sulphonated azo dyes in water and wastewater. TrAC Trends in Analytical Chemistry. 1997;**16**(7):405-419

[12] Mahendran V, Gogate PR. Degradation of acid scarlet 3R dye using oxidation strategies involving photocatalysis based on Fe doped TiO_2 photocatalyst, ultrasound and hydrogen peroxide. Separation and Purification Technology. 2021;**274**:119011

[13] Debnath S, Ballav N, Nyoni H, Maity A, Pillay K. Optimization and mechanism elucidation of the catalytic photo-degradation of the dyes eosin yellow (EY) and Naphthol blue black (NBB) by a polyaniline-coated titanium dioxide nanocomposite. Applied Catalysis B: Environmental. 2015;**163**: 330-342

[14] Perkampus HH. UV-VIS Spectroscopy and its Applications. Berlin, Heidelburg: Springer Science & Business Media; 8 Mar 2013

[15] Kumar A, Dixit U, Singh K, Gupta SP, Beg MS. Structure and properties of dyes and pigments. In: Dyes and Pigments-Novel Applications and Waste Treatment. Vol. 11. United Kingdom: IntechOpen Limited; 2021. p. 131

[16] Kurz W, Yetisen AK, Kaito MV, Koch AW. UV-sensitive wearable devices for colorimetric monitoring of UV exposure: Advanced Optical Materials. 2020;**8**(6):1901969 (1 of 9)

[17] Mansour R, Yusuf M. Natural dyes and pigments: Extraction and applications. In: Handbook of Renewable Materials for Coloration and Finishing. Vol. 9. United States: Scrivener Publishing LLC; 2018. pp. 75-102

[18] Karnjanawipagul P, Nittayanuntawech W, Rojsanga P, Suntornsuk L. Analysis of β-carotene in carrot by spectrophotometry. Mahidol University Journal of Pharmaceutical Science. 2010;**37**(1-2):8-16

[19] Januschewski E, Bischof G, Thanh BN, Bergmann P, Jerz G, Winterhalter P, et al. Rapid uv/Vis spectroscopic dye authentication assay for the determination and classification of reactive dyes, monascus pigments, and natural dyes in coloring foodstuff. Journal of Agricultural and Food Chemistry. 2020;**68**(42):11839-11845

Section 3

Applications of Colorimetric Analysis for Forensic and Defence Industry

Chapter 4

Forensic Analysis of Bloodstain Color

Tommy Bergmann and Dirk Labudde

Abstract

This book chapter delves into the field of colorimetric analysis of bloodstains in forensic science, focusing on its application in crime scene investigation. Therefore it provides a comprehensive overview of the biological background of age-induced color changes. The chapter begins with an introduction to the significance of blood evidence in solving crimes and the emergence of colorimetry as a valuable tool in blood analysis. The principles of forensic spectroscopy are explored, specifically its ability to provide information crucial to crime reconstruction, such as the age of bloodstains. The chapter discusses the transformation of hemoglobin derivatives over time and the corresponding measurable color changes that accompany aging blood traces.

Keywords: forensic science, blood pattern analysis, absorbance spectroscopy, bloodstain age estimation, influencing factors

1. Introduction

In the fascinating field of forensic science, the analysis of blood evidence is crucial in solving crimes. For centuries, the presence of blood at crime scenes has provided investigators with valuable clues to identify perpetrators and with information for the reconstruction of the crime scene with their sequence of events. With advances in analytical technology, colorimetry in particular has emerged as an invaluable tool in the forensically motivated examination of blood.

This chapter is devoted to colorimetric analysis of blood traces, a technique based on the absorption of light by blood components. We will dive deep into the principles of forensic spectroscopy, which makes it possible to obtain information relevant to crime reconstruction, such as the age of individual blood traces. It will be worked out to what extent the biological components of a blood trace are changed with aging and what measurable color changes accompany this.

We will examine the various components of blood responsible for color formation and address the challenges and limitations of this technique. A special focus will be placed on the manifold influencing factors of this colorimetric analysis, the understanding and mitigation of which is the content of current research.

IntechOpen

2. Spectroscopy in forensics

Whether in the structural elucidation of sample material found at the scene of a crime, in the determination of the origin of fibers and paints, in the detection of counterfeit drugs or banknotes, or in the age estimation of biological traces: spectroscopic methods are now established in almost all areas of forensic science. Spectroscopy is a biophysical measurement technique. It provides information about the size, shape, structure, charge, molecular weight, function, and dynamics of the macromolecules under investigation. Specific properties of light are exploited, allowing conclusions to be drawn about the state of the sample under investigation.

2.1 Term classification

Spectrometry or spectroscopy is a collection of physical methods in which the interactions of electromagnetic radiation with surrounding matter are observed and measured. The light can be reflected, transmitted, or absorbed. The interpretation is usually based on recorded spectra. These spectra are recorded with so-called spectrometers and are intensity distributions of radiation as a function of wavelength. They are used to identify substances and provide qualitative information about the quantitative composition of the sample. **Figure 1** shows a schematic picture of absorbance spectroscopy, which is one of these methods [1].

There is a wide range of spectroscopic examination methods in forensics that define or compare the color of trace materials to obtain information about their origin and to determine whether they were altered at the time of the crime. Examples include forensically motivated examinations of soil [3], vehicle paint [4, 5], textile fibers [6], postmortem interval [7], drugs [8], and others.

The spectroscopic method relevant to this chapter is spectroscopy in the ultraviolet and visible range (UV/VIS spectroscopy). It is an analysis method with which electrons are excited in the UV and visible range. It belongs to the invasive analysis methods. On the one hand, advantages are that only a small amount of sample is required, and on the other hand, a substrate-independent analysis is possible. In the case of organic substances, only unsaturated substances can be excited; thus, UV/VIS spectroscopy is a detection method for unsaturated, organic substances. An analysis of the recorded spectra enables making statements about different molecular bonds.

When recording a spectrum, the light coming from the light source is passed through a monochromator, which scans the set wavelength range. Since the visible and ultraviolet regions are being examined, two light sources are often used. These

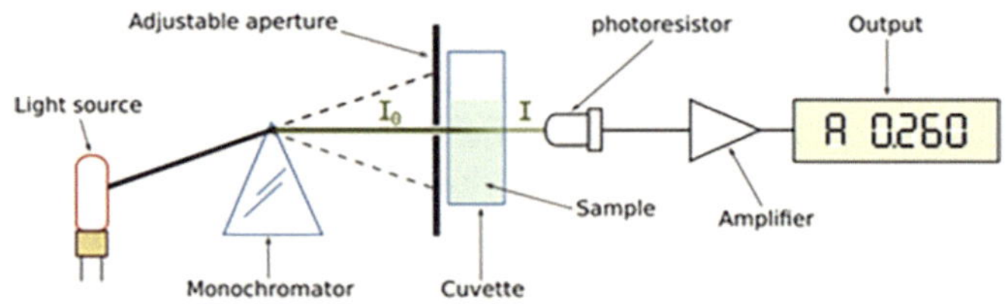

Figure 1.
Schematic setup of an absorbance spectroscope. Monochromatic light is brought to a fixed intensity Io with an aperture and passed through the sample. The remaining intensity is recorded, amplified, and can finally be read as absorbance value [1, 2].

are usually a tungsten filament lamp for the VIS range and a UV lamp for the other range, which alternate based on a preset wavelength.

So that a sample can be examined, so-called cuvettes are used. These contain the sample dissolved in a suitable solvent. The recording of the spectrum depends on the particular spectrophotometer. There are so-called single-beam spectrophotometers and double-beam spectrophotometers. The former first measures the pure solvent, followed by the measurement of the dissolved sample, which is corrected for the effect of the reference measurement. In double-beam spectrophotometers, the monochromatic light is split, allowing simultaneous measurement of sample and solvent. It is important that the solvent is also measured before each measurement so that the light from the sample and reference beams can be matched.

The absorbed radiation is converted to substance concentration using Lambert-Beer's law. This law gives the radiation intensity Eλ when passing through a medium with an absorbing substance as a function of the layer thickness d and the concentration of the absorbing substance c. The following formula applies:

$$E_{\lambda} = \varepsilon_{\lambda} \, c \, d \tag{1}$$

c: Mass concentration of the absorbing substance in liquid (in $\text{mol} \cdot \text{l}^{-1}$).

d: Layer thickness d of the irradiated body (in meters).

ελ: decadic extinction coefficient of the wavelength λ.

As mentioned before, in a UV/VIS spectrophotometer, electrons are excited by light. These are in the ground state and enter the excited state when they absorb light energy. The absorption of certain wavelengths depends on the molecular structure of the substance. The result is an absorption spectrum. Such a spectrum is called a measurement curve, which reflects the absorption of monochromatic light of different wavelengths of a substance [9].

2.2 Electromagnetic spectrum

In the wavelength range from about 380 nm (f = 789 THz) to 780 nm (f = 384 THz), electromagnetic radiation is visible to the human eye. It and sometimes also the adjacent ranges (infrared radiation and ultraviolet radiation), which can also be partially glimpsed due to the diffuse perception limits of the eye, are referred to as light. The actual nature of the underlying waves or particles was almost completely unknown until the beginning of the 19th century and is still debated today. In 1802 Thomas Young confirmed with his double-slit experiment Huygens' principle, which had already been designed in the 17th century. The opinion prevailing until then, that light can be regarded as a bundle of straight rays (ray optics or geometrical optics), was thus replaced by wave optics or physical optics, in which light is treated as a wave and can form wave fronts. However, the laws of geometrical order are still used today to represent relationships in an abstract way.

With the help of the wave-optical approach, properties such as diffraction, polarization, or color could be explained. Color, for example, depends on the wavelength or frequency of the light, which in turn depends on the speed of propagation of the wave in the respective medium. Two special forms are white light, a superposition of many waves, and monochromatic light, which has only one fixed wavelength/frequency. Known interactions of light with matter are, for example, reflection, scattering, refraction, or absorption. The wavelength λ is calculated from the phase velocity *c* divided by the frequency *f* [10].

3. Blood in forensic analysis and its parameters

Forensic examinations of bloodstains analyze their morphological appearance and variations in their composition. This requires a basic understanding of the physics, chemistry, and biology of blood, which is laid out in this section. Particular attention is paid to the blood color characteristics.

3.1 What is blood?

It is called the divine life element in many places, a symbol of the continuing vitality of the human body. Blood was considered the seat of the soul even in Egyptian times. Ritual ceremonies and healing practices with human or animal blood can be traced back to antiquity with the Mayas or in the Roman Empire and still take place today. In medicine, too, the so-called "red juice of life" plays a central role in anamnesis, since its components ensure the survivability of all body cells and also protect them from pathogens. It is easy to obtain, and any deviations in the composition of its components are important indicators of a present disease of the patient [11].

3.1.1 Some information about the anatomy and function

Not only is it proverbially thicker than water, and its average temperature of 38°C is about one degree warmer than body temperature (**Figure 2**). Its main function is to support and maintain various bodily functions by transporting substances. To do this, it is pumped through the blood vessels by the mechanical action of the heart, first away from the heart through the arteries and finally back to the heart through the veins. There is an average of about 70–80 ml of whole blood per kg of body weight in the human vascular system, which translates to about 5 to 6 liters for a 70 kg man. This whole blood consists of about 55 percent liquid blood plasma and 45 percent blood cells, which include platelets and red and white blood cells. The latter form the specific and nonspecific immune defense by detecting foreign substances (antibodies) and then binding or lysing them. These lymphocytes are of particular forensic

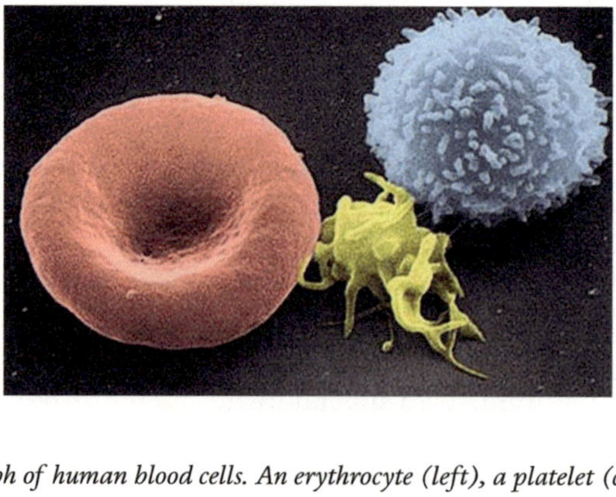

Figure 2.
Scanning electron micrograph of human blood cells. An erythrocyte (left), a platelet (center), and a leukocyte (right) can be seen [12].

relevance because they are the only type of blood cell that has a nucleus, making it possible to extract DNA for genetic fingerprinting. The much smaller blood platelets (thrombocytes), which make up only about one percent of all blood cells, perform the important task of blood clotting by attaching themselves to an injured vessel wall and thus forming a sealing blood clot. At around 96 percent, red blood cells (erythrocytes) make up the largest proportion of human blood in terms of quantity and mass. They transport the oxygen absorbed via the lungs to the organs, and the carbon dioxide accumulates there back to the lungs. Due to the lack of a nucleus, they look like small discs dented in the middle under the microscope [11].

3.2 The colors of the blood

About 90 percent of the dry mass of erythrocytes consists of the oxygen-binding protein hemoglobin. This iron-containing complex gives blood its red color (**Figure 3**). It consists of four subunits (2 "alpha"- and 2 "betachains"), which have a fold characteristic of globulins, in which an iron-II complex, the so-called heme, is bound. **Figure 4** contains a schematic representation of the ID 2H35 of the *protein data bank* (PDB). Under this ID, the theoretically calculated and experimentally proven structure of human hemoglobin can be found on the PDB. This freely accessible database contains information on the arrangement of amino acid chains (primary structure), their basic 3D arrangement (secondary structure), and the 3D structure of the whole chain (tertiary structure) of millions of proteins.

A central position in the heme complex is occupied by an iron atom, which can accept an oxygen molecule without changing its charge.

Within an organism (*in vivo*), hemoglobin is either in the oxygenated form (oxyhemoglobin) on its way from the lungs to the organs, or it is deoxygenated again on its way to the lungs (deoxyhemoglobin). If one oxygen molecule is bound to each of the four iron ions, the overall color changes from dark to light red; deoxyhemoglobin (unloaded) converts to oxyhemoglobin (oxygen-loaded). In rare cases, the already oxidized state undergoes further auto-oxidation, producing methemoglobin, which no longer has an affinity for oxygen. In humans, however, the enzyme methemoglobin reductase ensures that normally never more than 2 percent of the total hemoglobin is present in this stage since; otherwise, the organism's adequate oxygen supply would no longer be guaranteed.

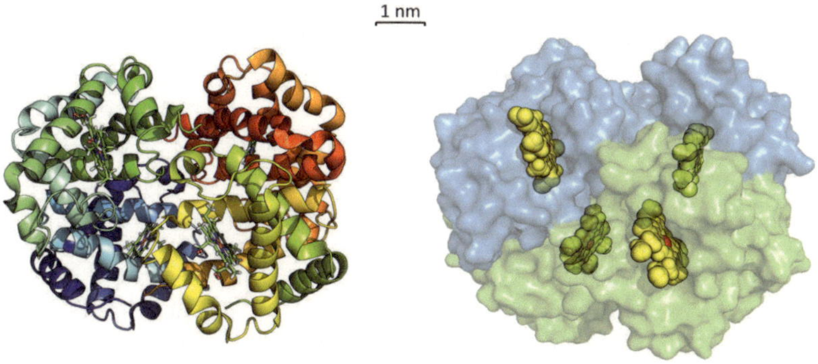

Figure 3.
Hemoglobin. PDB ID 2H35 with secondary structural elements color-highlighted (left) and heme groups with central iron atom schematically illustrated (right). The two colors (blue and green) on the right side represent the two different chains of HB.

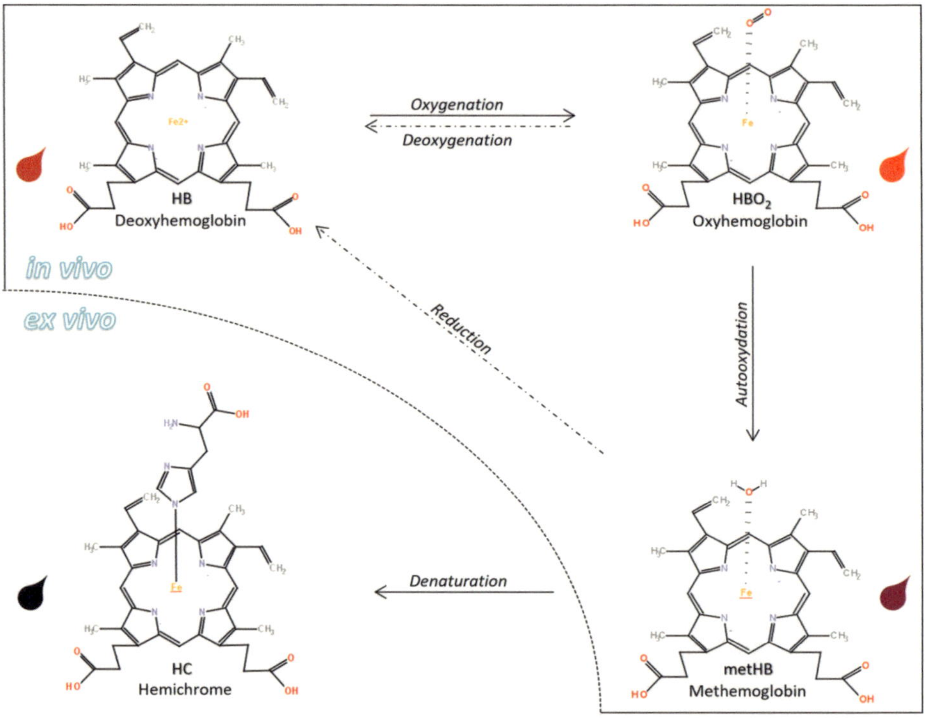

Figure 4.
Hemoglobin derivatives. The bivalent iron of hemoglobin (HBO_2) is oxidized to trivalent iron (metHB), which can no longer function as an oxygen transporter. In the subsequent denaturation, histidine complexes are often formed in which the iron is completely bound (HC). Due to the different peaks of these hemoglobin derivatives in the electromagnetic spectrum (see ***Figure 11****), the blood changes color from light red (HBO_2) to brown (metHB) to black (HC).*

The breakdown of old or damaged erythrocytes occurs for the most part in the spleen and liver. Since the deformability of aged or degenerated red blood cells is lower, they are increasingly trapped in the fine meshwork of the spleen, where they can be lysed. Certain scavenger cells of the liver (Browicz–Kupffer cells) can also take up the old cells from the bloodstream and break them down. During this process, the hemoglobin is broken down through several transformation processes to form urobilin (yellow color in urine) and stercobilin (brown color in feces) [14].

When blood is present outside the body (*ex vivo*), atmospheric oxygen oxidizes all hemoglobin deposits to oxyhemoglobin, which then slowly oxidizes further to methemoglobin. However, this process is irreversible due to the absence of methemoglobin reductase ex vivo. Thus, over time, methemoglobin increases at the same rate as oxyhemoglobin decreases.

The methemoglobin concentration increases over several days to a few weeks until the hemoglobin is completely oxidized. As decay progresses, amino acids (often histidine) become tightly attached to the central iron atom starting at two to 3 weeks, rendering the resulting hemichrome completely inactive (**Figure 5**). The temporal denaturation of the blood pigment is also accompanied by a color change. The initially bright red blood (oxyhemoglobin) gradually turns brownish (methemoglobin) until it finally appears almost black (hemichrome). This is, therefore, an oxidation of iron: the blood rusts. The exact process of blood decay ex vivo, unlike blood aging in vivo, has hardly been elucidated because it has no great medical relevance. For criminal

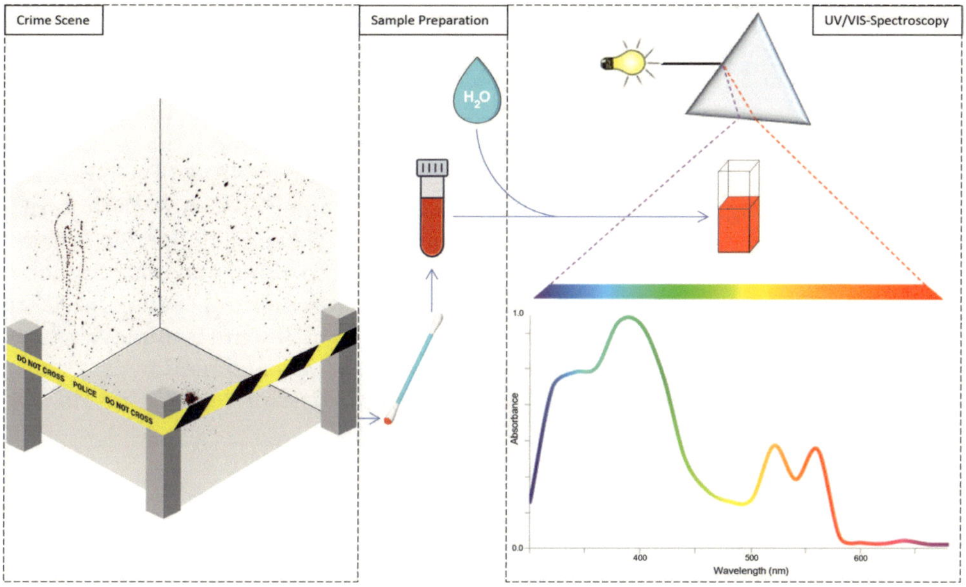

Figure 5.
Schematic overview of a UV/VIS spectroscopic analysis of bloodstains. Wet and dry traces of blood found at the crime scene are secured, for example, using cotton swabs. These samples are diluted with distilled water in the laboratory and transferred to a cuvette. This is then transilluminated with monochromatic light of different wavelengths, producing an image of how much light in which wavelength range is absorbed by the sample.

biology applications, there have been a few publications in recent decades in which the individual phases of blood aging ex vivo were examined in more detail. However, fresh blood was rarely used, or ethylenediaminetetraacetic acid (EDTA) was used as an anticoagulant, which means that the native circumstances can no longer be recreated too accurately. This will be discussed later in this chapter.

A drop of blood freshly discharged from the body thus undergoes a gradual change in composition and, consequently, in color. The speed of this change depends on external influencing factors such as surface properties, temperature, or humidity.

In simplified terms, the degradation of hemoglobin ex vivo takes place in a two-stage reaction, with the respective substeps having the rate constants k1 and k2 [15]:

$$Hb \rightarrow HbO_2 \xrightarrow{k_1} MetHb \xrightarrow{k_2} HC \qquad (2)$$

3.3 Spectrum of the blood

The method for forensic blood trace age estimation, which is particularly discussed in this chapter, is based on UV/VIS spectroscopy, a form of absorption spectroscopy (see **Figure 1**). In this process, the blood at the crime scene is picked up, for example, with a cotton swab and, diluted with distilled water, measured in the wavelength range between about 200 nm and 700 nm. **Figure 5** shows this process schematically.

As already mentioned, the denaturation of hemoglobin into its derivatives is accompanied by a color change (see Section 2.2). This change can also be followed in an absorbance spectrum of the blood, since the erythrocytes have a large share in it in terms of quantity. PATTERSON recognized that all hemoglobin derivatives have a strongly pronounced extinction band between about 400 and 425 nm, which is also called the Soret band [16]. Further spectroscopic experiments showed that HBO_2 has two more typical peaks at 542 and 577 nm. According to Zijlstra et al. the specific methHB peak is at 631.8 nm [17]. Thus, in a total spectrum of blood, the peaks of HBO_2 become weaker with time and give way to the peak of methHB (see **Figure 6**).

In addition, the significantly less studied and documented spectrum of HC is of interest. Absorption maxima are seen at wavelengths of 537 and 389 nm. Both maxima have the disadvantage of being overlapped by the absorption maxima of other hemoglobin derivatives in a whole blood spectrum. This maximum overlaps at 537 nm with that of HBO_2 at 542 nm. In addition, the maximum at 389 nm is immediately to the left of the Soret band and is influenced and possibly superimposed by it.

Figure 6 shows the effects of changes in hemoglobin derivatives over time in the whole blood spectrum.

4. Forensic significance

In criminal biology, blood traces have a special significance. They inevitably occur when the human body is violently impacted, for example, by blows or gunshots. The specific evaluation of the internal and external appearance of the stains allows a reconstruction of the crime and thus provides important information to convict

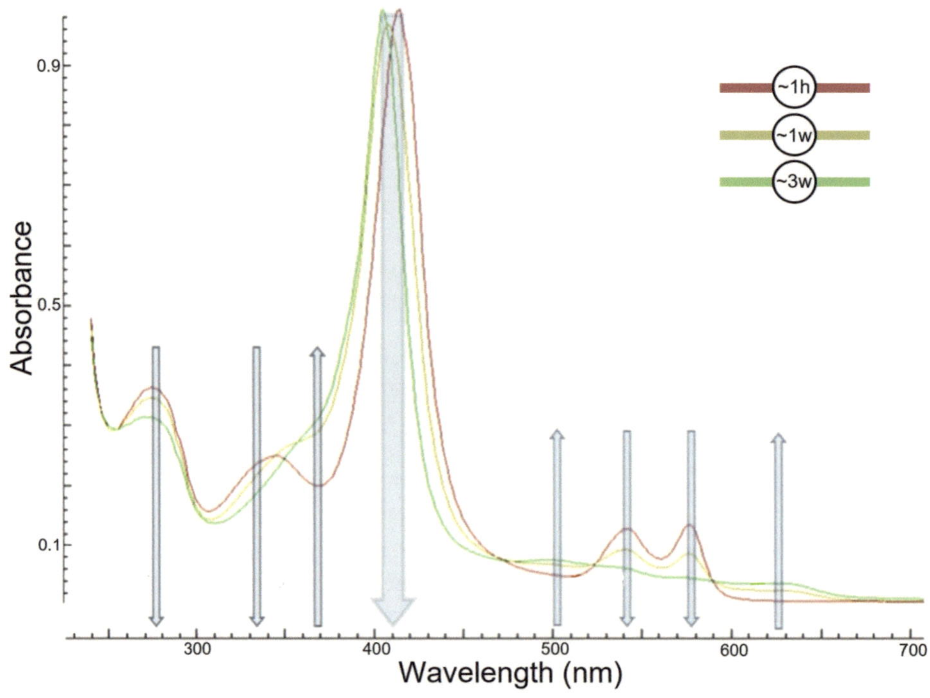

Figure 6.
Resulting blood spectrum. Comparison of spectra from three different age states of bloodstains (ca. 1d, ca. 1w, ca. 3w), color graded by age. Age-correlating changes in the spectrum are shown at different positions. They originate from different concentrations of the hemoglobin derivatives.

possible perpetrators. Forensic analysis of bloodstains can be roughly divided into bloodstain pattern analysis and blood droplet age determination. The methods of blood trace analysis attempt to reconstruct the general sequence of the crime, while blood drop age estimation attempts to narrow down the time of the crime.

4.1 Bloodstain pattern analysis

The morphological examination of a bloodstain often provides information about the direction, type, and intensity of the violent impact, the position of the persons involved in the room, and the distance to the source of the bleeding.

Historically, this way of examining a crime scene has existed since 1956, when the analysis of bloodstains was first mentioned by Lassaigne [18]. This was followed by Piotrowski [19] and Schmidtmann [20], both of whom attempted reconstruction using experimentally generated traces. In 1914, photographic documentation of traces for reconstruction was proposed for the first time by Ziemke [21]. Visualization of blood using 3-aminophthalic acid hydrazide (luminol) has been documented since 1930 [22]. Three years later, a paper on crown formation and the effect of different surfaces on blood stains appeared [23]. In 1939, Balthazard et al. discovered a relationship between the length and width of the stain to the angle of impact [24]. Macdonell postulated another correlation, that of the speed and flight distance of the tracks, in 1971 [25]. In 1975 Brinkmann dealt with the appearance of coughed-up tracks and the blood road crossings [26]. Backspatter traces in gunshot wounds, referred to as backspatter, were also addressed by Macdonell in 1982 [27]. In 1986, the physical background processes involved in the formation of bloodstains were studied for the first time by Pizzola et al. [28, 29]. The first computer-based approach was started by Carter from 1987 [30–32]. Since then, digital analysis systems existed, but routinely they are not used yet. Institutions dealing with bloodstain analysis are, e.g., the IABPA (International Association of Bloodstain Pattern Analysis) or the DGRM (German Society for Forensic Medicine).

Current research in this field is always investigating new physical occurrences that could possibly influence blood. An important requirement of current investigations is the continuous, accurate documentation or archiving of the found samples with high-resolution cameras held vertically over the stain [33]. In this way, one can still refer to this information in the later course of an investigation. The average blood droplet depends on the surface tension and has an average volume of about 50 µl [27], but ranges from 13 to 160 µl [34]. The smaller the volume, the higher the resulting gravitational acceleration. A liquid body always has the lowest surface energy with the largest possible volume/surface area ratio, which is why falling drops always have a slightly oscillating spherical shape or a spherical shape due to gravity [35]. If the trajectory of a formed spot is to be calculated, ballistic laws must be taken as a basis. Thus, Sellier and Kneubeuhl postulated that blood, like any other projectile, is a downward-opened parabola in flight due to the gravitational force of the earth. Typical flight ballistic quantities such as the air resistance or the rotational motion of the projectile (the droplet) must also be taken into account [36].

Since the resulting bloodstains of an act can never be completely identical, each case must, of course, be considered on its own merits. However, there are attempts to categorize or subdivide found bloodstains in order to bring some order to everyday investigation. Bloodstains are officially divided into contact traces and form traces. Contact traces, such as hair or handprints, had prior contact with a specific object that left a specific imprint. Inferences about the clothing worn or the presence of specific objects can be made when analyzing these prints. In the case of mold marks, the blood has traveled from a source to a specific location at a specific speed due to the application of a force. There, depending on the impact of the force, typical patterns are then created. Examples of mold marks are impact splash marks, drip marks, or coughed-up marks. Another common classification of blood traces is that of the DGRM. This distinguishes between passive traces, transfer traces, projected traces, and miscellaneous traces (see **Figure** 7).

4.2 Bloodstain age estimation

The exact temporal classification of a committed crime is essential in a forensic investigation. The techniques used for this purpose include, for example, rectal measurement of body temperature [38], examination of open wounds [39], or latent fingerprints [40]. Blood also changes in certain ways over time outside the body (see Section 3.2), so age determination is, in principle, possible. Over the last century, there have been many attempts to establish a precise age determination of blood traces. Due to the complexity of the examined source material blood, an enormous variety of starting points for this arose. Then as now, however, the methods were always coupled with a relatively high error rate. However, since blood is often the only trace of a crime that is found or can be used, the current aim is to overcome these inaccuracies.

4.2.1 Historical milestones

The first known mention of this subject in the literature was by the Italian Louis Tomellini in 1907 [41], who designed a color scale on which different age stages (up to 1 year) of a bloodstain could be seen. Three years later, LEERS explained that the

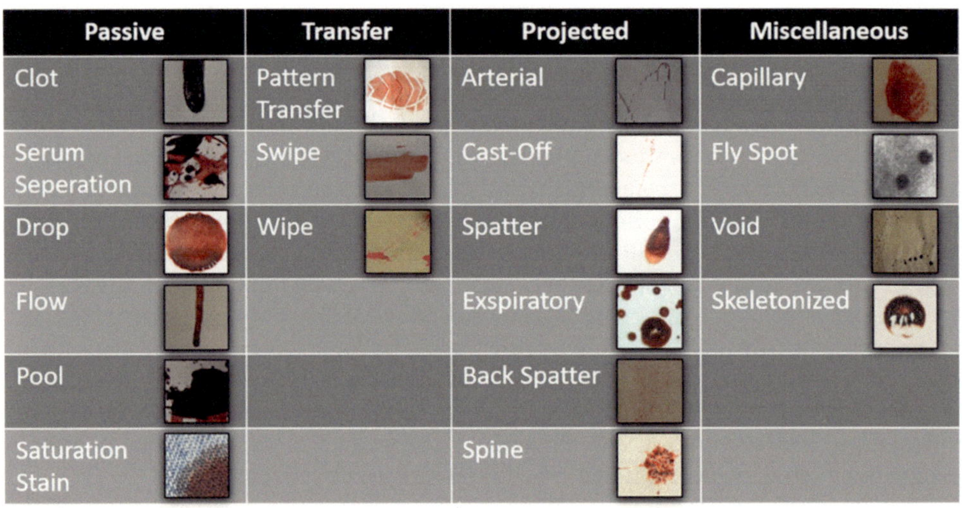

Figure 7.
Blood pattern classification. A schematic depiction of one of the most common classification schemes for bloodstain patterns in Europe. It distinguishes between traces that were generated passively ("passive"), traces that were created actively through transfer ("transfer"), and traces that were created actively through projection ("projected"). Other traces, which are difficult to assign, are assigned to the category "miscellaneous." (mod. From [37]).

spectrum of blood in the visible range changes significantly. The reason for this was the temporal change of hemoglobin [42]. Schwarzacher took a different approach in 1930 and investigated the dissolution rates of old bloodstains, which became progressively more difficult to solubilize [43]. Next, in 1937, the peroxidase activity of hemoglobin was studied using a Guaiacum-based assay. The color response of this experiment was directly correlated with time [44]. The chemical background of the occurring color change during blood aging was investigated and broken down in 1950 [45]. Ten years later, there were the first spectrometric studies showing that the changes were environmentally dependent [16]. The foundation for this was laid by Ziemke as early as 1901 when he studied various hemoglobin derivatives and determined their absorption bands [46]. This methodology was further elaborated and improved by Kleihauer et al. [47] and Patterson et al. [16]. In 1972, Nuorteva proved that blood age determination is also possible with the help of entomology [48]. Bloodstained objects or pieces of clothing are visited by flies like corpses. Their larvae subsequently reveal something about the colonization period.

4.2.2 Current methods of forensic blood age estimation

The list of current techniques for blood age determination is long. There are approaches with several methods for almost all blood components. Due to this diversity, the approaches are subdivided below according to the blood component under investigation.

Since the predominantly clear, liquid portion of blood, blood serum, contains some specific proteins, their observation lends itself to detecting any signs of degradation over time. For example, changes in the various globulins (α-,β-, and γ-globulin) can reveal something about the age of the stain under study [49]. In this immunoeletrophoretic approach, the β- and γ-globulins decayed much more slowly, which served as an approach to calculating age. UV absorption photometry can further be used to quantify the activity of various enzymes (lactate dehydrogenase, glutamate oxaloacetate transaminase, and glutamate pyruvate transaminase), yielding a specific image for each time point as the individual enzymes decompose at different rates [48]. A relatively new possibility for age determination using blood plasma is the use of different degradation rates of contained amino acids. This is accomplished by aspartic acid racemization, which has already been successfully used for rough age estimation of fossils [50]. In 2010, Ackermann et al. created a completely new idea with their ELISA-based (Enzyme-linked-immunosorbent-assay) approach, which, however, answers a different question. With the help of two hormones (melatonin and cortisol), which have a fixed and resolved level in the blood at any time of the day, it is possible to find out the time of the day when the blood spot develops but not the actual age [51].

Since white blood cells, in contrast to red blood cells, have a cell nucleus, so it is possible to extract and analyze RNA here. This is quantified by fluorescence spectroscopy. The ratio of mRNA to rRNA increases slowly and can be converted to ex vivo blood age [52]. This technique can be used for very long postmortem periods, and thus has no upper time limit [53].

4.3 Current colorimetric methods for blood age estimation

The majority of blood age determination methods are based on changes in red blood cells (erythrocytes) and their constituents over time. A frequently used technology for this is the so-called high-speed liquid chromatography (HPLC), in which the smallest components of liquids can be separated from each other. Hemoglobin and its derivatives or degradation products can be easily detected and distinguished by this equipment. Age correlation using the resulting peaks of the chromoproteins has already been achieved but was not particularly accurate [54]. The peak of a previously unassigned protein (called protein X) allows a better age correlation. The absorbance peaks of this protein, which increase with age, do not lie between 250 and 600 nm, which means that a relationship with hemoglobin can be excluded [55]. The presence of other possibly relevant, as yet unknown proteins (previously named protein Y and Z) was noted but not further investigated. Since the central iron atoms of oxygenated (Oxyhämoglobin→ $Fe2^{+}$) and oxidized hemoglobin (e.g., Methämoglobin→ $Fe3^{+}$) are differently charged and thus have different numbers of free orbitals, they also respond differently to an applied magnetic field. Such fields can be generated with the aid of so-called electron spin resonance spectrometers. Examined blood traces give four specific signals in it, which stand for the different degeneration products of the red blood pigment [56]. The oxyhemoglobin itself has no signal in this process because it is diamagnetic. Due to unexpected values, such as the decrease in the hemichrome peak after about 3 weeks, no further research has been done in this area. Since the concentration of freely accessible (unoxidized oxygen) decreases sharply, especially in the first hours, with the decrease of oxyhemoglobin and increase of methemoglobin, Matsuoka et al. tried to deduce an age relationship from this [57]. In this experiment, a Clark electrode (oxygen electrode) measured the decrease in oxygen atoms in the first 20 hours after blood application. A significant decrease was measured, but it was strongly temperature dependent. Another possibility to examine red blood cells is the so-called atomic force microscopy. In this complex, very high-resolution (down to the nanometer range) technique, the surface of a sample is scanned line by line with an oscillating leaf spring (cantilever). This determines the structure but also the elasticity of the surface. Especially the elasticity of erythrocytes is a helpful parameter here, since it increases enormously in ex vivo stored blood during the first 30 days (from ca.40 kPa to ca. 2.5 GPa) [58]. The fact that the actual structure of the cells hardly changes during this period and that the error rate of age estimation is uncomfortably high also made this research area stand still at first. Since the most noticeable visible change of an aging bloodstain is the transition of the color from red to brown, it seems reasonable to check the absorbance in the visible range (about 390 to 790 nm) for changes. Accordingly, there are some approaches in which this is done with a reflectance spectrometer that records the intensity of reflected rays from a white monochromatic light source. For example, Kind et al. showed that the two typical peaks of oxygen of oxyhemoglobin decrease at 540 and 576 nm and allow estimation of blood age [59].

The rather diffuse Soret band of a blood sample (which is the large peak at about 414 nm) also changes significantly [60]. Imaging of the entire visible region over time provides insight into the decrease or increase in the individual degradation products of hemoglobin and thus also allows for age correlation [61]. In addition to imaging the visible region, peaks arising in the near-infrared region can also help with this problem. NIR scanning can also be used in other contexts with blood, namely to clearly separate it from other blood-like substances [62]. Since 2009, there is now the possibility to investigate the large band between 1350 and 1900 nm. Here, although there is a rapid and highly visible decrease, this is largely due to the decrease of water in the blood, and it subsides after only a few hours. In contrast, the actual protein bands between 1900 and 2500 nm show little diversity over time [63]. In addition to the qualitative evaluation of spectra, quantitative measurements can also be performed at a specific wavelength range to determine the concentration of an ingredient. A recently published article by Agudelo et al. uses this approach to perform age estimation on both young and older blood samples [64]. The content of this publication is the use of a bipartite bioaffinity assay. Two biomarkers (creatine kinase and alanine transaminase) were thereby added to the buffered blood serum. Both ultimately lead to the build-up of pyruvates, which in turn promote the degradation of nicotinamide adenine dinucleotide hybrid (NADH) to nicotinamide adenine dinucleotide (NAD^+). Measurement of nicotinamide adenine dinucleotide consumption occurs at 340 nm in real time. Because the two biomarkers have different but well-defined denaturation times, the creatine kinase portion of the assay can more accurately date younger blood, and the alanine transaminase portion can more accurately date older blood. An approach, also recently published, allows age determination in the field with the aid of computer-assisted image processing [65]. Here, a cell phone camera is placed on a type of specially constructed darkroom, and the stain to be analyzed is imaged from above. The obtained images are then used to train a classification algorithm (Random Forests). Henceforth, the trained algorithm is to distinguish by age all blood images taken in the same way. Thanakiatkrai et al. not only tested the suitability of different smartphones for this experiment but also simulated different environmental conditions.

In the last 10 years, there have been further developments in almost all areas of spectroscopic bloodstain age estimation. The existing approaches in reflectance spectroscopy have been optimized [66] and adapted so that even the background color no longer produces large measurement errors [67]. Age estimation via NIR has also been optimized so that imaging is now mobile [68]. In addition to method optimization, there has also been further investigation with respect to changes in blood composition *ex vivo* over time. This area has paid particular attention to studies of changes in diverse red blood cell constituents [69–71] to further understand their influence on spectral examination methods. The information on morphological changes, which has also been studied, reveals how much the external appearance depends on environmental conditions [72].

4.4 Influencing factors

Now that some techniques have been presented, the question arises as to which influencing factors alter blood aging and in what way. Such influences are, e.g., solar radiation, humidity, ambient temperature, type, and condition of the soil or also the pretreatment of blood with anticoagulants. All these influences can affect the drying out of a blood trace, which is why it brings with it a certain uncertainty in terms of age estimation. Fresh blood traces are usually still liquid and not completely dried out. In principle, it is already possible to use the determination of blood age for crime detection. Blood dries faster ex vivo than in the body of a corpse.

4.4.1 Temperature, light exposure and humidity

As mentioned in Section 2, HbO_2 in blood exposed to atmospheric oxygen denatures to metHb and later to hemichrome. By analyzing spectroscopically prepared data, Bremmer et al. concluded that the first reaction is strongly temperature dependent: the higher the temperature, the faster HbO_2 is converted to metHb. Humidity has no influence on the process. The transition from metHb to HC, however, is strongly dependent on both temperature and humidity [15].

The dependencies of air humidity and light exposure are shown as examples in **Figure 8**.

Hanson and Ballantyne investigated another aspect of blood aging in 2010. As part of their research, they recognized that in a spectroscopic examination of aging blood traces, the Soret band of hemoglobin appears to migrate over time. They referred to this phenomenon of leftward shift as blue shift. They found that this shift is strongly

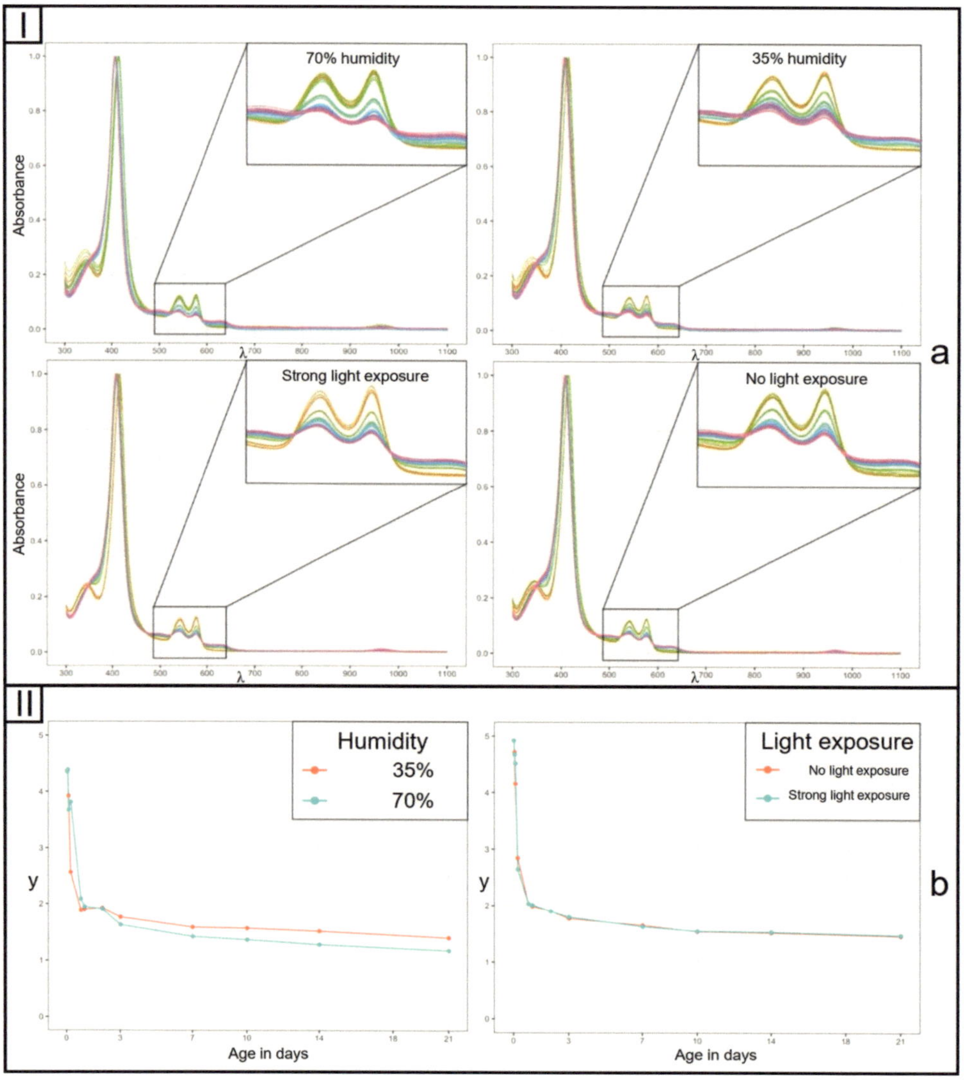

Figure 8.
Influence of light exposure and humidity on bloodstain aging. I] within 3 weeks, blood spectra change due to free atmospheric oxygen ex vivo. This change is influenced by prevailing humidity (a) and solar radiation (b), among other factors. II] a concentration-independent ratio was formed from the major oxy- and methemoglobin peaks (P1 = 542 nm, P2 = 578 nm, P3 = 500 nm, P4 = 532 nm) using the formula y = (P1 + P2)/(P3 + P4). It is noticeable that higher values of humidity favor faster blood aging (a) while there is no significant effect of light intensity on the conversion of HbO_2 to metHB (b).

dependent on both temperature and humidity. The higher the humidity or the storage temperature, the lower the value by which the Soret peak shifts [60].

Figure 9 shows the shift in the Soret band and its dependence on ambient temperature.

4.4.2 Surface

In 2017, we found that different surfaces (cotton, glass, polyester) show a noticeable influence on spectroscopic results [73]. In the context of investigations carried out in the laboratory, it is therefore advisable to always use the same substrate for

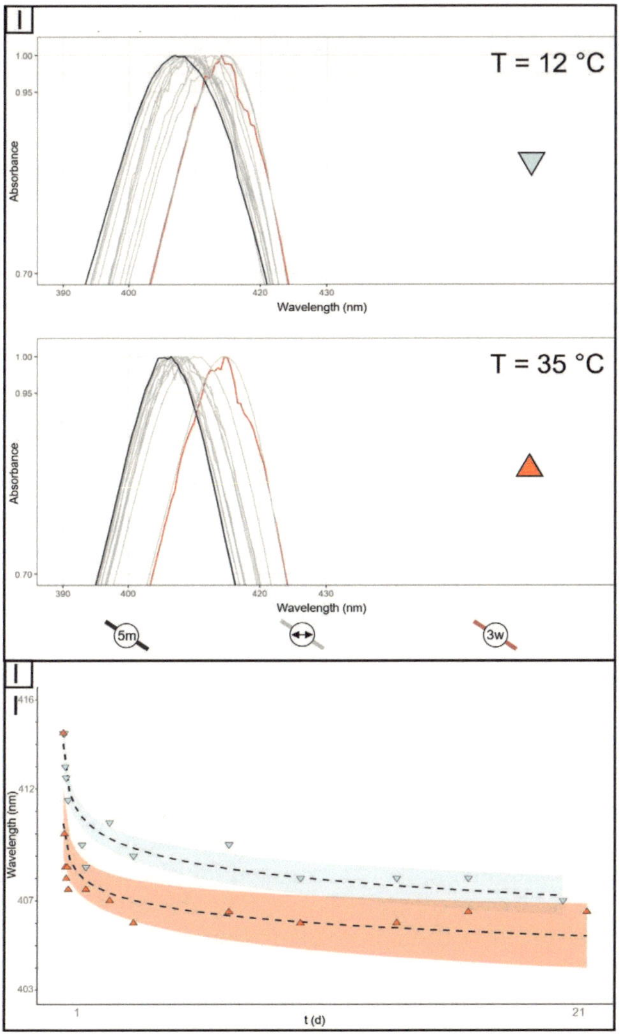

Figure 9.
Temperature dependence of the Soret shift. I] the so-called "blue shift" of the soret band within 3 weeks at 12°C (a) and 35°C (b) and a relative humidity of 35% in each case can be seen. II] from the main oxy- and methemoglobin peaks (P1 = 542 nm, P2 = 578 nm, P3 = 500 nm, P4 = 632 nm) a concentration-independent ratio was formed according to the formula y = (P1 + P2)/(P3 + P4). The differences in the diagram show that higher humidity causes faster "blue drift" or blood aging.

experiments on blood aging. When evaluating case-relevant traces, however, it should always be remembered that blood trace aging proceeds differently on different surfaces.

4.4.3 Anticoagulants

Anticoagulants are frequently used in medicine to prevent potentially dangerous complications during surgery or to reduce the formation of clots in the circulation. Anticoagulants are also popular for obtaining valuable data sets in a forensic context (e.g., generating different blood patterns or recreating crime scenes), as the associated delayed clotting allows blood samples to be used for longer periods of time.

There are different types of anticoagulants such as citrate, hirudin or *ethylenediaminetetraacetic acid* (EDTA), each of which has an influence on different subprocesses of blood coagulation. In 2020, we showed that at least one of them (EDTA) alters the aging of blood in a way that spectroscopic methods for age estimation give false results [74]. This is logical and is also mostly taken into account when evaluating crime-relevant blood traces, but the creation of data sets in the laboratory to recreate the crime or to research the influencing factors discussed here are still frequently carried out with EDTA-amended blood samples for the advantages just described. This should be prevented in the future because otherwise untreated blood drops from the crime scene are compared with EDTA-treated blood drops from previously st6att found experiments.

5. Conclusions

Colorimetric analysis methods have a firm place in forensics. In the specific field of blood trace pattern analysis, they are mainly used for age estimation. The transformation of hemoglobin derivatives observed in this process has been known for over 100 years, but even today, the methods based on it are being optimized to obtain ever more accurate results under all assumed circumstances of the crime. So far, it has been understood that blood color reflects not only information about health status and possible intoxications but also the time the blood has been ex vivo. This is due to the omnipresent atmospheric oxygen ex vivo, which alters the ratio of the hemoglobin derivatives, observable in the wavelength range of about 400 nm - 600 nm. Under real conditions at the scene, the recorded spectral data can be affected by various environmental conditions. In order to correctly interpret the color-coded information, these influences must be investigated and taken into account. The more these external influences are understood and accounted for, the more reliable the blood age estimate will be. For this reason, current scientific publications in this field predominantly deal with optimized ways of calculating the influence of environmental conditions on blood aging rather than completely new methods.

What the future holds is, of course, uncertain, but it is likely that the trend of other forensic disciplines will continue, and parts of blood trace pattern analysis will also be automated using computer-aided methods. Approaches to this have already been made and range from AI-based analysis methods to the so-called "Smart Forensic Phone," a mobile detection system based on a cell phone [65]. Such approaches are not intended to overhaul existing methods but to facilitate their use in everyday forensics. An app under development (not yet published by us) demonstrates this (see **Figure 10**). A clear and easy-to-use interface on a standard cell phone

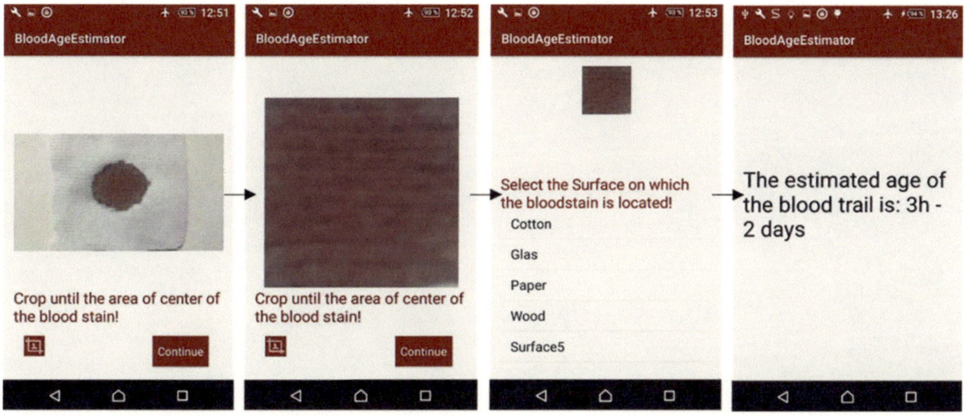

Figure 10.
"Blood age estimator" app. User interface of an application under development for user-friendly recording and age estimation of blood traces.

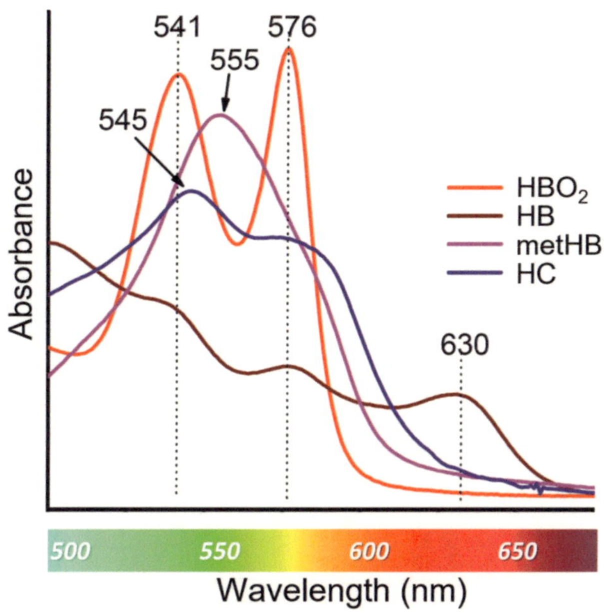

Figure 11.
Hem derivative spectra. Absorbance values of HBO_2 (red), metHB (brown), HB (pink), and HC (blue) in the range from 450 to 700 nm (mod. From [13]).

allows the use of colorimetric methods for forensic blood trace age estimation at almost any location by nontrained personnel.

Conflict of interest

The authors declare no conflict of interest.

Author details

Tommy Bergmann* and Dirk Labudde
Forensic Science Investigation Lab (FoSIL), University of Applied Sciences
Mittweida, Germany

*Address all correspondence to: bergmann@hsmw.de

References

[1] Schmidt W. Optische Spektroskopie: Eine Einführung für Naturwissenschaftler und Techniker. Weinheim: VCH Verlagsgesellschaft mbH; 1994

[2] Yassine Mrabet. Essential parts of a spectrophotometer [Internet]. 2008. Available from: https://commons.wikimedia.org/wiki/File:Spetrophotometer-en.svg [Accessed: May 15, 2023].

[3] Dong CE, Webb JB, Bottrell MC, Saginor I, Lee BD, Stern LA. Strengths, limitations, and recommendations for instrumental color measurement in forensic soil characterization. Journal of Forensic Sciences. 2020;**65**(2):438-449. DOI: 10.1111/1556-4029.14193 Epub 2019 Sep 16

[4] Nogayeva S, Gooch J, Frascione N. The forensic investigation of vehicle–pedestrian collisions: A review. Science & Justice. 2021;**61**(2):112-118

[5] Duarte JM, Sales NGS, Sousa MH, Bridge C, Maric M, de Andrade Gomes J. Automotive paint analysis: How far has science advanced in the last ten years? TrAC Trends in Analytical Chemistry. 2020;**132**:116061

[6] Goodpaster JV, Liszewski EA. Forensic analysis of dyed textile fibers. Analytical and Bioanalytical Chemistry. 2009;**394**:2009-2018. DOI: 10.1007/s00216-009-2885-7

[7] Vanezis P. Assessing hypostasis by colorimetry. Forensic Science International. 1991;**52**(1):1-3

[8] Namera A, Nakamoto A, Saito T, et al. Colorimetric detection and chromatographic analyses of designer drugs in biological materials: A comprehensive review. Forensic Toxicology. 2011;**29**:1-24. DOI: 10.1007/s11419-010-0107-9

[9] Böcker J. Spektroskopie - Kapitel 1. In: Einführung. 1.Auflage ed. Vogel Buchverlag; 1997. pp. 26-61

[10] Lehmann T. UV/VIS-Spektroskopie. Berlin: Freie Universität; 1997-2009

[11] Schmidt RF, Thews G. Physiologie des Menschen. Springer; 1997. pp. 411-435-604-606

[12] Jakob Suckale. Red_White_Blood_cells [Internet]. 2016. Available from: https://commons.wikimedia.org/wiki/File:Red_White_Blood_cells.jpg [Accessed: May 15, 2023]

[13] Meng F, Alayash AI. Determination of extinction coefficients of human hemoglobin in various redox states. Analytical Biochemistry. 2017;**521**:11-19. DOI: 10.1016/j.ab.2017.01.002

[14] Braun F, Legrun S. Spezielle Toxikologie für Chemiker. Teubner, S138 ff; 1999

[15] Bremmer RH, De Bruin DM, De Joode M, Jan Buma W, Van Leeuwen T, Aalders MCG. Biphasic oxidation of oxy-hemoglobin in bloodstains. PLOS One. 2011;**6**(7):e21845. DOI: 10.1371/journal.pone.0021845

[16] Patterson D. Use of reflectance measurements in assessing the colour changes of ageing bloodstains. Nature. 1960;**187**:688-689. DOI: 10.1038/187688a0

[17] Wg Z, Buursma A, Van Ow A. Visible and near Infrared Absorption Spectra of Human and Animal Haemoglobin. Zeist: VSP BV; 2000

[18] Lassaigne J.B., Neue Untersuchungen von zur Erkennung

von Blutflecken auf Eisen und Stahl, Vierteljahrsschrift für gerichtliche und öffentliche Medicin, Bände 9-10, S. 285, 1856

[19] Piotrowski E. Über Entstehung der Blutspuren nach Hiebwunden des Kopfes. Virchows Jahresber. 1895;**I**:449. DOI: 10.7396/2020_1_C

[20] Schmidtmann A. Handbuch der gerichtlichen Medizin. Hirschwald Verlag; 1905

[21] Ziemke E. Die Untersuchung von Blutspuren. In: Lochte T, editor. Gerichtsärztliche und polizeiärztliche Technik. Wiesbaden: Bergmann; 1914. pp. 152-215

[22] Lt L. Hedgecock dg: Chemiluminescence in the visualization of forensic bloodstains. Journal of Forensic Sciences. 1978;**23**:550-555

[23] Lochte T. Über die Kronenbildung des auffallenden Bluttropfes und ihre Beziehung zu sekundären Blutspritzern. Dtsch Z ges Gerichtl Med. 1933;**22**:387-397

[24] Baltharzard V, Prèdelièvre R, Desoille H, Dèrobert L. Etude Des Gouttes De Sang Projetè. In: Presented at the 22nd Congress of Forensic Medicine of the French Language, Directeur de L'Institut de Médecine Légale Université René Descartes, Professeur a. Paris: Hadengue; 1939. pp. 265-323

[25] Hl M. Flight Characteristics and Stain Patterns of Human Blood. Washington DC: U.S. Department of Justice, L.E.A.A., National Institute of Law Enforcement and Criminal Justice (Nile CIPR71-74); 1971. DOI: 10.1002/9780470061589.fsa066

[26] Brinkmann B. Expertisen an biologischen Spuren - Bestandsaufnahme, zukünftige Trends. Z. Rechtsmedizin. 1988;**100**:39-54. DOI: 10.1007/BF00200363

[27] Hl M. Bloodstain Pattern Interpretation. Corning New York: Laboratory of Forensic Science; 1982. DOI: 10.1007/BF00200363

[28] Pa P, Roth S. De Forest Pr: Blood droplet dynamics -I. Journal of Forensic Sciences. 1986;**31**(1):36-49

[29] Pa P, Roth S. De Forest Pr: Blood droplet dynamics -II. Journal of Forensic Sciences. 1986;**31**(2):50-64

[30] Al C. The directional analysis of bloodstain patterns - theory and experimental validation. Journal of the Canadian Society of Forensic Science. 2001;**34**(4):173-189. DOI: 10.1080/00085030.2001.10757527

[31] Al C. Directional analysis of bloodstain patterns with a computer. In: James SH, Kish P, Sutton E, Paulette T, editors. Principles of Bloodstiain Pattern Analysis: Theory and Practice. Boca Raton Florida: Taylor & Francis-CRC Press; 2001. pp. 241-261. DOI: 10.1080/00085030.2001.10757527

[32] Carter A et al. Further validation of the BackTrack™ computer program for bloodstain pattern analysis – Precision and accuracy. IABPA News. 2005;**21**(3):15-22 NCJ Number 213441

[33] Tl W. Documentation of bloodstain pattern evidence. The Journal of Forensic Identification. 1995;**45**(4):396-408

[34] Tl L. Diameter of bloodstains as function of origin, distance fallen and volume of drop. IABPA News. 1985;**2**:12-16

[35] Courtney M. Blood Spatter Evidence and Reconstruction of Events. Fort

Worth TX: Forensic Consultant Services; 1991

[36] Und SK, Bp K. Wundballistik und ihre ballistischen Grundlagen. Berlin: Springer; 2001

[37] Bevel T. Bloodstain Pattern Analysis: With an Introduction to Crime Scene Reconstruction. Boca Raton: CRC Press; 2002

[38] Kaliszan M, Hauser R, Kernbach-Wighton G. Estimation of the time of death based on the assessment of post mortem processes with emphasis on body cooling. Legal Medicine (Tokyo, Japan). 2009;**11**(3):111-117. DOI: 10.1016/j.legalmed.2008.12.002 Epub 2009

[39] Grellner W, Madea B. Demands on scientific studies: Vitality of wounds and wound age estimation. Forensic Science International. 2007;**165**:150-154. DOI: 10.1016/j.forsciint.2006.05.029

[40] Wertheim K. Fingerprint age determination: Is there any hope? The Journal of Forensic Identification. 2002;**53**:42-49 NCJ Number 198540

[41] Tomellini L. De l'emplol d'une table chromatiqie pour les taches du sang. Arch. d'Antropologie criminelle de Criminol. 1907;**14**

[42] Leers O. Die forensische Blutuntersuchung. Berlin: Springer; 1910

[43] Schwarzacher. Determination of the age of bloodstains. The American Journal of Police Science. 1930;**1**. DOI: 10.2307/1147182

[44] Schwarz F. Quantitative untersuchungen der katalase und peroxydase im blutfleck. International Journal of Legal Medicine. 1937;**27**:1-34

[45] Rauschke J. Beitrag zur Frage der Alterbestimmung von Blutspuren. International Journal of Legal Medicine. 1951;**S.40**:578-584. DOI: 10.1007/BF01347087

[46] Ziemke FM. Beiträge zur Spektroskopie des Blutes. Arch. f. Anat. u. Physiol. 1901. DOI: 10.1007/BF01347087

[47] Kleihauer E, Stein G, Schmidt G. Beitrag zur alterbestimmung von blutflecken. Archiv für Kriminologie. 1967;**140**:84-94

[48] Nuorteva P. Age determination of a blood stain in a decaying shirt by entomological means. Forensic Science. 1974;**3**:89-94. DOI: 10.1016/0300-9432(74)90012-0

[49] Rajamannar K. Determination of the age of bloodstains using immunoelectrophoresis. Journal of Forensic Sciences. 1977;**22**

[50] Tsutsumi A, Yamamoto Y, Ishizu H. Determination of the age bloodstains by enzyme activities in blood cells. Japanese Journal of Legal Medicine. 1983;**37**:6

[51] Arany S, Ohtani S. Age estimation of bloodstains: A preliminary report based on aspartic acid racemization rate. Forensic Science International. 2011;**212**(1-3):e36-e39. DOI: 10.1016/j.forsciint.2011.05.015

[52] Ackermann K, Ballantyne KN, Kayser M. Estimating trace deposition time with circadian biomarkers: A prospective and versatile tool for crime scene reconstruction. International Journal of Legal Medicine. 2010;**124**(5):387-395. DOI: 10.1016/j.forsciint.2011.05.015

[53] Anderson S, Howard B, Hobbs GR, Bishop CP. A method for determining the age of a bloodstain. Forensic Science International. 2005;**148**(1):37-45. DOI: 10.1016/j.forsciint.2004.04.071

[54] Inoue H, Takabe F, Iwasa M, Maeno Y. Identification of fetal hemoglobin and simultaneous estimation of bloodstain age by high-performance liquid chromatography. International Journal of Legal Medicine. 1991;**104**(3):127-131. DOI: 10.1007/BF01369715

[55] Inoue H, Takabe F, Iwasa M, Maeno Y, Seko Y. A new marker for estimation of bloodstain age by high performance liquid chromatography. Forensic Science International. 1992;**57**:17-27. DOI: 10.1016/0379-0738(92)90041-t

[56] Miki T, Kai A, Ikeya M. Electron spin resonance of bloodstains and its application to the estimation of time after bleeding. Forensic Science International. 1987;**35**:149-158. DOI: 10.1016/0379-0738(87)90051-X

[57] Matsuoka T, Taguchi T, Okuda J. Estimation of bloodstain age by rapid determinations of oxyhemoglobin by use of oxygen-electrode and total hemoglobin. Biological & Pharmaceutical Bulletin. 1995;**18**:1031-1035. DOI: 10.1248/bpb.18.1031

[58] Strasser S, Zink A, Kada G, Hinterdorfer P, Peschel O, Heckl WM, et al. Age determination of blood spots in forensic medicine by force spectroscopy. Forensic Science International. 2007;**170**:8-14. DOI: 10.1016/j.forsciint.2006.08.023

[59] Kind SS, Patterson D, Owen GW. Estimation of the age of dried blood stains by spectrophotometric method. Forensic Science. 1972;**1**:27-54. DOI: 10.1016/0300-9432(72)90146-X

[60] Hanson EK, Ballantyne J. A blue spectral shift of the hemoglobin soret band correlates with the age (time since deposition) of dried bloodstains. PLoS One. 2010;**5**:e12830. DOI: 10.1371/journal.pone.0012830

[61] Bremmer RH, Nadort A, van Gemert MJC, van Leeuwen TG, Aalders MC. Age estimation of blood stains by hemoglobin derivative determination using reflection spectroscopy. Forensic Science International. 2010;**206**(1-3):166-171. DOI: 10.1016/j.forsciint.2010.07.034

[62] Brooke H, Baranowski MR, McCutcheon JN, Morgan SL, Myrick ML. Multimode imaging in the thermal infrared for chemical contrast enhancement. Part 1: Methodology. Analytical Chemistry. 2010;**82**:8412-8420. DOI: 10.1021/ac101109w

[63] Botonjic-Sehic E, Brown CW, Lamontagne M, Tsaparikos M. Forensic application of near-infrared spectroscopy: Aging of bloodstains. Spectroscopy. 2009;**24**:42-48 1939-1900

[64] Agudelo J, Huynh C, Halámek J. Forensic determination of blood sample age using a bioaffinity-based assay. Analyst. 2015;**140**:1411. DOI: 10.1039/C4AN02269F

[65] Thanakiatkrai P, Yaodam A, Kitpipit T. Age estimation of bloodstains using smartphones and digital image analysis. Forensic Science International. 2013;**233**:288-297. DOI: 10.1016/j.forsciint.2013.09.027

[66] Edelman GJ, Roos M, Bolck A, Aalders MC. Practical implementation of blood stain age estimation using spectroscopy. IEEE Journal of Selected Topics in Quantum Electronics. 2016, 7200107;**22**(3):415-421. DOI: 10.1109/JSTQE.2016.2536655

[67] Sun H et al. Accurate age estimation of bloodstains based on

visible reflectance spectroscopy and Chemometrics methods. IEEE Photonics Journal. 2017, 6500614;**9**(1):1-14. DOI: 10.1109/JPHOT.2017.2651580

[68] Pereira JFQ, Silva CS, Maria JL, Vieira MF, Pimentel AB, Honorato RS. Evaluation and identification of blood stains with handheld NIR spectrometer. Microchemical Journal. 2017;**133**:561-566. DOI: 10.1016/j.microc.2017.04.038

[69] Sakota D, Kani Y, Kosaka R, Nishida M, Maruyama O. Simultaneous determination of hemolysis and hematocrit in extracorporeal circulation by plasma surface reflectance spectroscopy. Annual International Conference of the IEEE Engineering in Medicine and Biology Society. 2013;**2013**:6764-6767. DOI: 10.1109/EMBC.2013.6611109

[70] Lavrinenko IA, Vashanov GA, Artyukhov VG. Decomposition of hemoglobin UV absorption spectrum into absorption spectra of prosthetic group and apoprotein by means of an additive model. Biofizika. 2015;**60**(2):253-261 Russian

[71] Varlet V, Ryser E, Augsburger M, Palmiere C. Stability of postmortem methemoglobin: Artifactual changes caused by storage conditions. Forensic Science International. 2018;**283**:21-28. DOI: 10.1016/j.forsciint.2017.12.009 Epub 2017 Dec 8

[72] De Wael K, Lepot L. Morphological details in bloodstain particles. Forensic Science International. 2015;**246**:50-54. DOI: 10.1016/j.forsciint.2014.10.033 Epub 2014 Nov 4

[73] Bergmann T, Heinke F, Labudde D. Towards substrate-independent age estimation of blood stains based on dimensionality reduction and k-nearest neighbor classification of absorbance spectroscopic data. Forensic Science International. 2017;**278**:1-8. DOI: 10.1016/j.forsciint.2017.05.023 Epub 2017 Jun 8

[74] Bergmann T, Leberecht C, Labudde D. Analysis of the influence of EDTA-treated reference samples on forensic bloodstain age estimation. Forensic Science International. 2021;**325**:110876. DOI: 10.1016/j.forsciint.2021.110876

Section 4

Applications of Colorimetric Analysis for Food Industry

Chapter 5

Application of Colorimetry in Food Industries

Kakoli Dutta and Rosalin Nath

Abstract

The acceptance of any food product be it raw, cooked or processed is first evaluated by its color, flavor and texture. Human visual perception cannot accurately measure a particular color intensity, it may vary due to various factors. Though visual color standards and assessment are used in food industries but with the recent advancement of technology the quality assessment procedure is shifting towards colorimetric analysis. Colorimetry is the technology used for color measurement. There are various techniques and color models used in colorimetry while in the food sector the CIE LAB and RGB color model is mainly used as it is the closest to the mechanism of a human eye. Color measurement is a complex subject and the consumer acceptability of a food product, post-harvest management and market statistics depends on it. This chapter provides a brief discussion about the type of colorants, importance of color in the food industry, different color scales used in colorimetry and the various applications.

Keywords: colorimetry, CIEL*a*b*, RGB, color model, color space

1. Introduction

As long as a person has good eyesight colors can be perceived, but it is not possible to quantify colors with sound vision [1]. Visual color judgments are affected by various factors, such as lighting conditions, angle of observation, color blindness etc. The only consistent and subjective was of color quality control is to measure it by a standardized instrument [2]. The quantitative measurement of colors can be done by colorimetry. The science of color measurement is known as colorimetry. Colorimetry is the technology used for measuring a wide range of food product including fresh as well as processed food items. Colorimetry can be also used to measure qualitative and quantitative measurement of components like amino-acid, sugar, enzymes, polyphenols, antioxidants etc.

Food quality generally depends upon three major parameters- nutritional quality, sensory acceptability and safety measures. Sensory acceptability includes attributes like color, flavor, aroma, body texture, mouthfeel and cultural appropriateness [3]. When it comes to customer acceptability color is the most important parameter on basis of which the quality of the food material is judged [4]. Color is considered as the most important physical attribute of food; it is demonstrated as the indicator of physical chemical and sensorial product quality [4, 5]. Food color can be assessed using either the human visual system or color measuring instruments. Comparing colored references in regulated lighting is a requirement for visual system. Color standards are frequently utilized as reference materials to carry out color analysis in this method. This method implies slower inspection and requires proper training and that is the reason it is recommendable to employ color measurement through instruments. The most used instruments to measure color are colorimeter such as Hunterlab colorimeter, Dr. Lange colorimeter, Minolta chroma meter.

2. Classification of food colors

The very first sensory attribute by which a food is judged is its color. Human senses are somehow trained to perceive food with certain colors to be fresh and if the color deviates from its expectation, the food is rejected. Food with esthetic color and flavor are likely to be accepted as nutritious and hence consumed. Natural pigments which are associated with fruits and vegetables are usually bright and vivid. During processing of foods these pigments often undergoes physical and chemical changes causing color degradation and thus reduces the esthetic appeal. This is the reason food colorants both natural and artificial are used to restore the original appearance [6]. To ensure color uniformity, to help preserve the visual indication and to provide an attractive appeal to food colors are used [7]. Both artificial and natural colors play an important role in enhancing the color intensity, storage and quality control. However, there are strict government guidelines on the usage limit of these colors to reduce the toxicological effect if any and to ensure the health and safety of the consumers.

2.1 Natural

Natural raw materials when used as a source for extracting coloring components are usually known as Natural colors. The examples include anthocyanin from red grapes, betanine from beet root, chlorophyll from spinach or any green leafy vegetables, cucurmin from turmeric, lycopene from tomatoes, capsorubin from paprika and so on [8]. Often organic solvents are used to extract these natural and water-insoluble colors like carotene from carrots. To enhance the pigment stability against oxidation certain antioxidants like tocopherols are used. Certain plant which are not used for human consumptions are used for color extraction such as lutein from marigold flower, bixin or annatto from seeds of *Bixa orellano*, chlorophyll from grass [9, 10].

Other non-food raw sources from where natural colors are extracted are carminic acid which is orange in color from cochineal, carotenoid from nonedible algae, blue color pigment from spirulina, genipin from *Gardenia jasminoides* and lycopene from fungal biomass [11]. Caramel is a natural color but it does not occur in nature, when edible carbohydrates or sugars are heated at high temperature in presence of acid, alkali or salts [12]. Sources and physical properties of few natural colors has been mentioned below.

2.1.1 Anthocyanins

Anthocyanins belongs to the group of compounds called flavonoid; it is a water-soluble pigment. Six aglycone anthocyanidins found in foods are namely pelargonidin, delphinidin, malvidin, cyanidin, petunidin and peonidin (**Figure 1**). Anthocyanins are

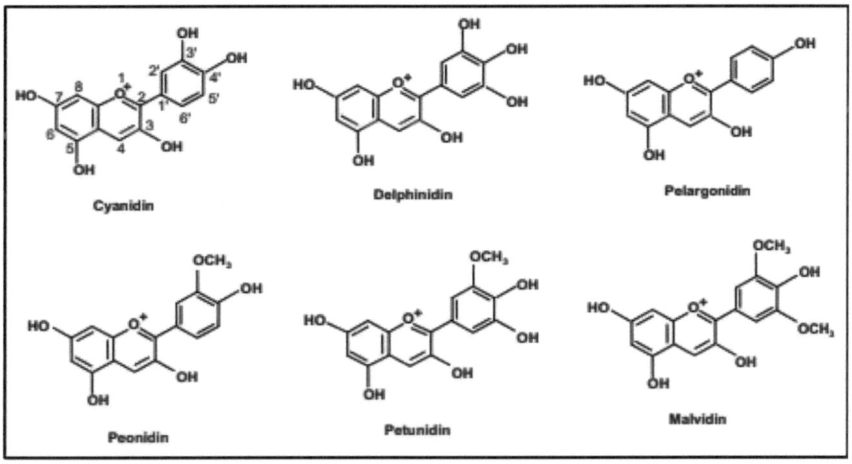

Figure 1.
Six most commonly found anthocyanins in foods [13].

mainly found in fruits such as red/purple grapes, black currant, cherry, raspberry, cranberry, strawberry, blueberries, red/purple cabbages and so on [14, 15].

The factors effecting the color and stability of the color are the chemical structure, light, temperature, oxygen, enzymes, metallic ions and other phenolics. Use of anthocyanins as a colorant has its limitations due to its low stability and interaction with other constituents in the food matrix [13].

2.1.2 Betalains

Betalains are nitrogen containing, water-soluble pigments ranging from red to violet betacyanins and yellow to orange betaxanthins. It is synthesized from tyrosine which is an amino acid into betacyanin and betaxanthin. Betalamic acid structure is the common chromophore among all the betalain pigments [16]. These pigments absorb visible radiation from 476 to 600 nm with 537 nm being the optimum at pH 5.0. The color shift (bathochromic) from betacyanins to betaxanthins, that is from 50 to 70 nm occurs due to the aromatic rings of cyclo-3-(3,4-dihydroxyphenylalanine). Beetroot is considered to be the most common source for the extraction of the food colorant betanin. Besides there are other sources such as Malabar Spinach (*Basella rubra*), tuber of *Ullucus tuberosus*, leaves and grains Amaranthas, dragon fruit, cactus fruit etc. [17, 18]. The stability of Betalains depends upon pigment content, degree of glucosylation, water activity, pH, antioxidants, temperature and nitrogen in atmosphere [19].

2.1.3 Carotenoids

Carotenoids are pigments that is soluble in fat, the color of the pigment ranges from yellow-orange-red usually found in plants and some animal. Carotenoids can be widely categorized into carotenes and xanthophyl. Carotenoids mostly present in foods are α and β carotene, lycopene, lutein, β cryptoxanthin and xanthophyll [20]. Sources of β-cryptoxanthin, α and β-carotene are carrot, oranges, sweet potato, mango, apricot, green leafy vegetables, tangerine, pumpkins and palm fruits [21]. Sources of lycopene are tomatoes, pink fleshed guavas, red-papayas and watermelon.

Corn, pumpkin and butternut squash are the sources of xanthophyll [22]. The stability of carotenoid may be affected by processing (juicing, puree formation, high temp cooking), disruption of food matrix (peeling, shredding, slicing), storage conditions (presence of light, atmosphere), water activity, enzymes etc. [23].

2.2 Artificial

Artificial food colors also known as synthetic food colors are the chemical modification of specific precursor compounds. The sources can be both natural and chemical; for example, indigo carmine is a synthetic color but obtained by sulfonation of indigo which is a natural source, whereas azo dyes providing various shades of red, yellow blue or black are produced by complete chemical synthesis of aromatic amine [24]. Artificial colors have advantages over natural colors with respect to heat or light sensitivity, color stability and chemical interactions. Artificial color has high color intensity and also do not impart any flavor or taste of its own. The colors that can be used in food is authorized by the European Commission (EC) in Regulation. The colors list in **Table 1** are the permitted synthetic colors to be used in food.

2.2.1 Tartrazine

Tartrazine (E 102) is a yellow color pigment; it is a water-soluble mono-azo color. This color remains stable in acidic condition, it can withstand temperature as high as 200°C [26]. Tartrazine is made by diazotization of 4-amino benzenesulfonic acid along with 4,5-dihydro-5-oxo-1-(4-sulfophenyl)-1H-pyrazole-3-carboxylic acid or its esters (**Figure 2a**). The color was isolated as potassium, calcium and sodium salts [27]. The Acceptable Daily Intake (ADI) limit of tartrazine is 7.5 mg/kg body weight per day. It has been found to be significantly harmful for asthma patients and even children at a higher dose and that is the reason this color has been banned in few countries [28]. The absorption of tartrazine in human body is as low as 5% and rest of the fraction passes undigested and unabsorbed through urine via the kidney.

Synthetic food colors	
E-number	**Name**
E 102	Tartrazine
E 110	Sunset Yellow FCF/Orange Yellow S
E 122	Azorubine, Carmoisine
E 123	Amaranth
E 124	Ponceau 4R, Cochineal Red A
E 129	Allura Red AC
E 151	Brilliant Black BN, Black PN
E 155	Brown HT
E 180	Litholrubine BK

Table 1.
Authorized azo-colors according to European Commission regulation [25].

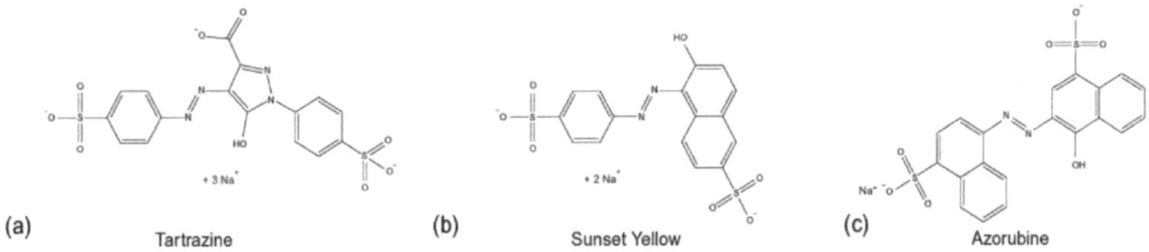

Figure 2.
Structural formula of azo-dyes—(a) Tartrazine, (b) sunset yellow and (c) Azorubine.

2.2.2 Sunset yellow

Sunset yellow (E 110) is a petroleum based yellow to orange azo dye for coloring food with the chemical name disodium 2-hydroxy-1-(4-sulfonatophenylazo) naphthalene-6-sulfonate (**Figure 2b**). Ca and K salts of this dye are permitted to be used as food color with a max permissible limit of 500 mg/kg used in sauces, coatings, salami and sausages, 50 mg/kg in baked and confectionary item, 50 mg/l is beverages and sometimes in various cheese and cheese rinds [29]. According to Joint FAO/WHO Expert Committee on Food Additives (JECFA) and Scientific Committee on Food of the European Commission (SCF) the ADI of sunset yellow is 0–4 mg/kg [27].

2.2.3 Azorubine/Carmosine

Azorubine (E122) is a red color, its chemical name is disodium 4-hydroxy-3-(4-sulfonato-1-naphthylazo) naphthalene-1-sulfonate, **Figure 2c** shows the structure. The maximum permissible limit of this dye is 500 mg/kg in preparation of sauces, coatings, salmon substitutes, processed meat or 50 mg/l in case of non-alcoholic or other fruit-based beverages. This color can also be used in minimum amount in edible cheese rind and casings. According to WHO the ADI of azorubine is 0–4 mg/kg [30], reportedly it does not have any carcinogenic effects, it may cause allergy in rare instances to The WHO and FAO approved dosages [31].

3. Importance of color measurement

Over the past few decades, the overall living standard of people have improved which has led to an increase in the awareness of quality of food products. As the demand for food has increased, the rate of food adulteration and food fraud have also increased leading to health risk exposure [32]. This is the reason food quality assessment is an essential step to be followed to ensure public health and safety [33]. There are various parameters of the assessment such as appearance (size, shape, color), texture and smell. Among these attributes food color is considered to be the primary parameter to evaluate food quality [34, 35]. The visual monitoring is the most convenient inspection method. Human eye is capable of recognizing almost 1million different colors [36] but due to poor color memory human brain cannot recall the exact color intensity observed previously [37]. Visual color monitoring depends upon a lot of complex factors such as lighting, human perception, illumination angles and so on. Therefore, instrumental methods are considered to be optimum.

Color may be defined as the observation experienced by a person when the radiation from the visible spectrum (400–800 nm) comes in contact with the eye [38]. Whereas, colorant is the pigment which give the product its color. Three important elements required for the phenomena of color to manifest are—a colored item, a source of light in the visible spectrum and a person/observer. White light can be absorbed, reflected or scattered when it collides with the object to be viewed. The main factor which determines the color of the object is the selective absorption of a particular light wavelengths [39].

4. Color specification system

The color estimation system can be of two types: Visual system and Instrumental methods.

4.1 Visual system (Munsell color system)

The best known visual color-ordering system was developed by A.H. Munsell in the year 1905 and thus known as Munsell system (**Figure 3**). In this system, there are 5 primary color such as red, yellow, green, blue and purple and five secondary pairs green yellow, red purple, blue green, purple blue and yellow red. These secondary colors are known as Hue. When the color quality is described by its lightness or darkness that is from white to gray and black it is known as Value. The Value is measured in the scale of 0 which is absolute black to 10 which is absolute white. Chroma describes the saturation of each color, for example addition of pure blue hue to a gray until it reaches original blue. This point is the saturation where there will be no more change in the shade [42].

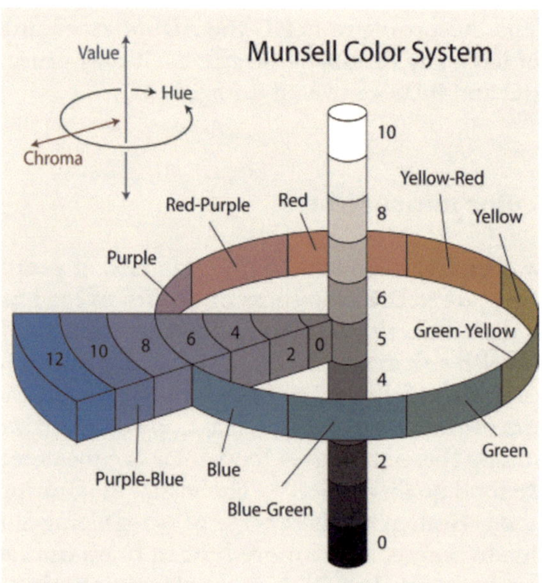

Figure 3.
Munsell visual system (source: [40, 41]).

4.2 Instrumental methods (CIE color system)

CIE color system is a well-known system to describe color, CIE (Commission Internationale de l'Eclairage) is an important and main international body concerned with color measurement. In most food industries, the color measurement instruments work on color-space system: L*, a*, b* defined by CIE. This trichromatic system can match any color by combination of the three primary color that is red, green and blue (RGB) represented by X, Y and Z axis respectively. The CIE system uses the chromaticity diagram to identify a particular color (**Figure 4**). Uppercase X, Y and Z gives the tristimulus values while lowercase x, y and z represent red, green and blue respectively. The value of x can be calculated by the formula, $x = X/(X + Y + Z)$ and the value of y and z can be obtained by substituting the value of x.

4.2.1 Color models and color spaces

Color models are the visualization of multidimensional models depicting the color spectrum.

RGB color model uses three primary colors namely red, green and blue which are varied at specific amount to produce a desired color. This model depicts the mapping of the red, green and blue colors as x, y and z axis. Combinations of the primary colors produces secondary color (**Figure 5**). The common applications of RGB model are Cathode ray tube, LCD, LED or a large monitor. In RGB color model there is a sensor which records the intensity of red, green and blue spectrum in each pixel. In case of food RGB coordinates are transformed into CIE L*a*b* color space for improved calculation of the colors [43].

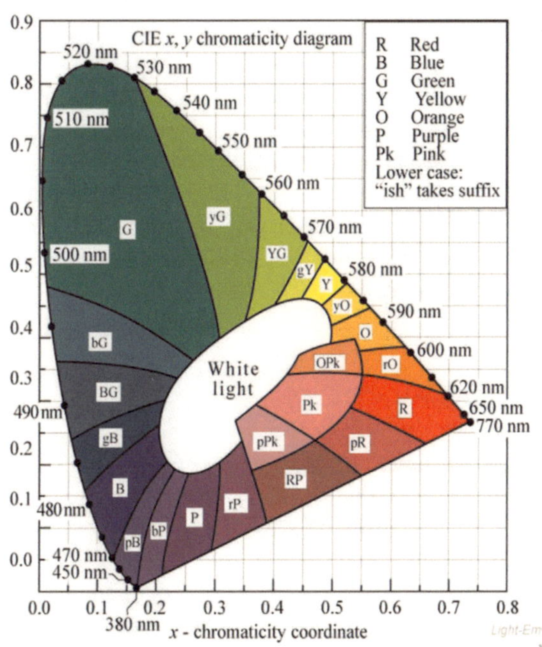

Figure 4.
CIE chromaticity diagram.

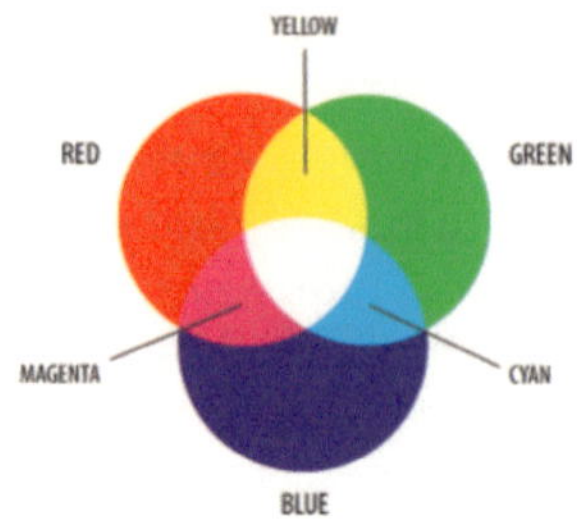

Figure 5.
Secondary colors from RGB.

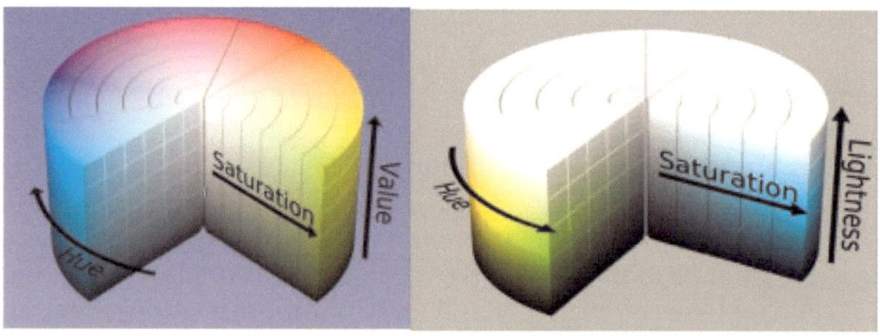

Figure 6.
The cylindrical diagram of HSV [41, 44, 45].

CMYK and CMY has the primary color Cyan, Magenta and Yellow and the "K" stands for the key. This is a device dependent model and the working is similar to the RGB color model. The cyan, magenta, yellow and the black key pigments transferred to white surface and subtract other color from the surface to finally get the product color.

HSV color model has three components. Hue, Saturation and Value integrated to form a geometry of a cylinder. Often Value is replaced by 'brightness' or 'lightness' called HSB or HSL respectively (**Figure 6**). Both are common cylindrical representation in the RGB color model. The angle around the cylinder passing through the centre is the measurement of hue. The distance from the central axis to the boundary measures the saturation. The difference of color along the height of the central axis is the Value, Lightness or Brightness [46].

5. Principle of colorimeter

When a beam of incident light (I_0) travels through a specific solution, some of the light is reflected (I_r), some of it is absorbed (I_a), and the remaining light is transmitted (I_t). This is the basis for the principle of a colorimeter.

Therefore, $I_0 = I_r + I_a + I_t$.

The two fundamental rules of photometry that form the foundation of the colorimeter concept can be used to define this mathematical relationship between the amount of light absorbed and the material concentration.

According to Beer's Law, the concentration of the solute in the solution is directly proportional to the amount of light absorbed.

DOI: http://dx.doi.org/10.5772/intechopen.112099

$$\mathrm{Log}_{10}\ \mathrm{I}_0/\mathrm{I}_t = a_s c$$

Where, a_s = absorbency index; c = concentration of solution.

According to the Lambert's law, the length and thickness of the solution under examination directly affect how much light is absorbed.

$$A = \log_{10}\ \mathrm{I}_0/\mathrm{I}_t = a_s b$$

Where, A = Absorbance of test; a_s = Absorbance of standard; b = length/thickness of the solution.

The combined mathematical expression of Beer-Lambert Law is:

$$\mathrm{Log}_{10}\ \mathrm{I}_0/\mathrm{I}_t = a_s bc$$

If b is kept constant by taking a cuvette or a standard cell, then:

$$\mathrm{Log}_{10}\ \mathrm{I}_0/\mathrm{I}_t = a_s c$$

Where Absorption index is given by: a_s = A/cl.

In the combined mathematical expression of Beer-Lambert's Law,

A $\rightarrow$ Absorbance or optical density of the solution.

c $\rightarrow$ Concentration of the absorbing material (gm/lit).

l $\rightarrow$ distance traveled by light in solution (cm).

In simple terms, the combined principle of Beer-Lambert's law states that the amount of light absorbed by a color solution is directly proportional to the concentration of the solution and the length of the light path through the solution, $A \propto cl$.

Thus, A = $\in$cl

Where, $\in \rightarrow$ Absorption Coefficient

Colorimeter is used to measure concentration of a colored solution or compound. It works in the visible spectrum of light ranging from 400 to 800 nm. As discussed, the working principle is based on Beer-Lambert law. Any colored chemical compound taken as a sample in the cuvette absorbs some light and the amount of light absorbed is proportional to the concentration of that that particular compound in the sample. The filament lamp which is the source of light inside the colorimeter emits light and it passes through the cuvette made of glass with a fixed thickness. Portion of the light is absorbed and the rest of the transmitted light falls on the photoelectric detectors. These detectors measure the intensity of incident light and the amount of absorbed light is calculated (**Figure** 7).

6. Colorimeters used in food industries

Colorimeters usually measures the primary sources which emits light and secondary sources which reflects external light.

6.1 Tristimulus colorimeter

The tristimulus values are obtained optically and not mathematically as the apparatus has been created based on the cone cells present in the retina of a human eye. The three filters present in the instrument, functions like the three types of cone cells

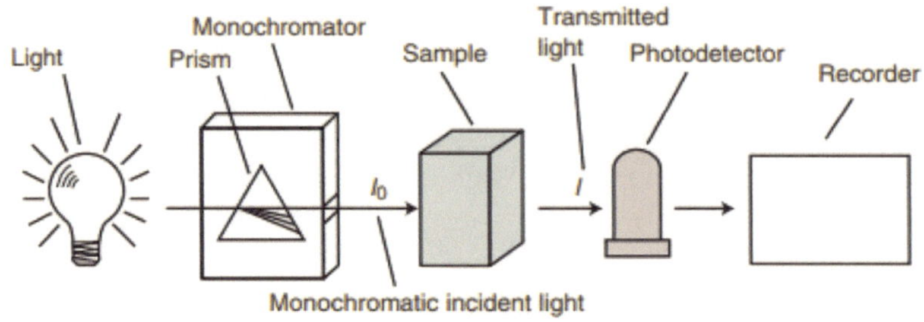

Figure 7.
Basic components of colorimeter [47].

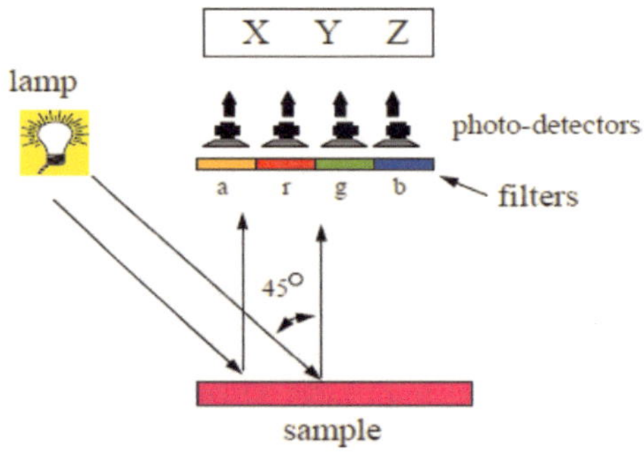

Figure 8.
Tristimulus colorimeter [49].

in human eye. A tristimulus colorimeter is made of three main components: (1) a light source, (2) color filters which can modify the incident or reflected light's energy distribution and (3) a photoelectric detector which can convert reflected ray into electrical output [48]. The measurement done on this colorimeter is usually comparative and that is the reason it is important to calibrate the standards using similar colors that is to be measured. **Figure 8** shows the schematic diagram of a 4 filter tristimulus colorimeter.

6.2 Spectrophotometer

Spectrophotometer measures the absorbance of light when an incident light passes through the colored sample solution. The basic principle of spectrophotometer almost similar to the colorimeter except that spectrophotometer can also measure the absorbance in visible (400–800 nm), UV region (200–400 nm) and Infrared region (700–1000 nm) of the electromagnetic spectrum.

There are two types of spectrophotometers: single beam and double beam.

In case of single beam spectrophotometer, the absorption of the blank sample (reference sample) is measured and then in place of blank the sample to be analyzed is placed and the absorption of light is measured. The absorption value of the blank sample is subtracted from the absorption value of main sample to get the result.

In case of double beam spectrophotometer, the source of illumination source gets split into two beams after the light passes through the lens. One beam of light passes through the blank sample while the other passes through the original sample. Both the transmitted rays hit a photo detector and then it is recombined to reach the

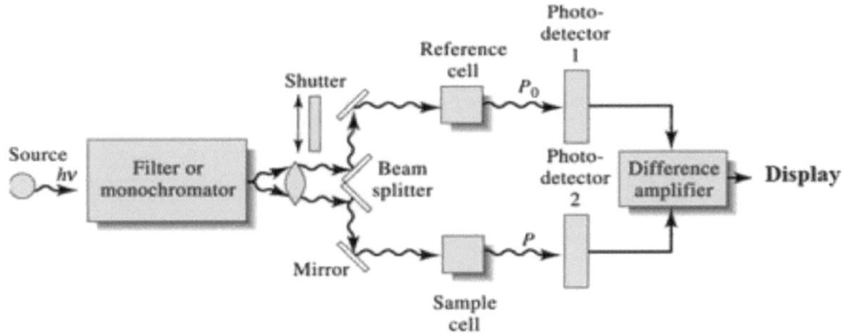

Figure 9.
Double beam spectrophotometer.

monochromator. The display section gives the reading by subtracting the blank sample from the original sample. A schematic diagram in **Figure 9** shows the double beam spectrophotometer. The cuvettes holding the sample are made of quartz and the light sources used are deuterium lamps for UV measurement, tungsten lamp for visible range. In recent advancement xenon arc lamps has come into use for both UV and visible spectrum.

6.3 Spectrocolorimeter

The basic principle of Spectrocolorimeter is quite similar to a spectrophotometer. Spectrocolorimeter are a type of colorimeter used to measure the color across the spectrum, it can analyze the reflectance and transmittance of the sample for wavelength in visible spectrum. It can provide the colorimetric data as CIE L*a*b* values or X, Y, and Z tristimulus values [50].

6.4 Densitometer

Densitometer actually measure the color of any particular food or any other substance based on its density. The color perception of any substance to be measured may change depending upon how dense the substance it. The color intensity of a single thin layer of any object differs from a thick film or layer. This type of colorimeter measures the amount of red, blue and green object that gets reflected from the object. Densitometer uses the optical filters of red, green and blue and measures the transmission of cyan, magenta and yellow.

6.5 Computer or machine vision

Computer vision came into existence around 1970s when it was used as mimicking the human vision system in the field of photography, camera schematics and projections. After a few decades later around 2010 this technique was used in other possible hi-end tasks like machine learning, graphics, fingerprint recognition, face recognition, image processing and so on [51]. The computer vision system setup consists of a light source or an illumination device, a camera to process the image objects, a frame grabber for scanning the image, a computer or a microprocessor to store the images and the computation based on a specific application and lastly a

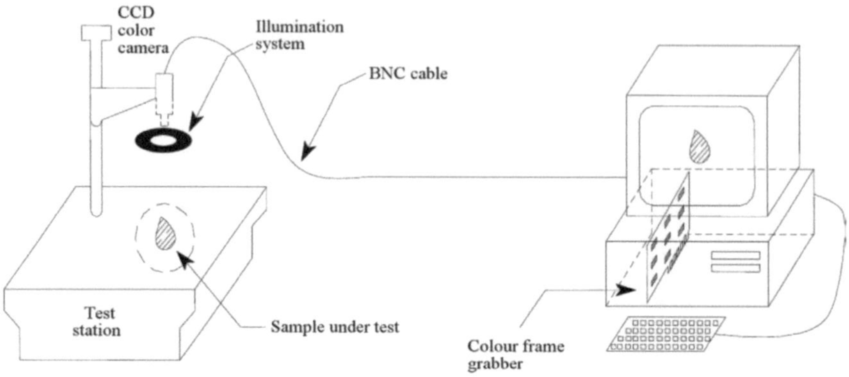

Figure 10.
Essential parts of a typical computer vision system [52].

high-resolution monitor to see the image [52]. **Figure 10** shows the schematic diagram of a typical computer vision system setup.

This non-destructive method was found to be used in food industry for color grading, sorting, maturity, and freshness of fruits and vegetables [53]. Image processing algorithms has to be made based on parameters such as shape, color homogeneity, defects of any types of fruits or vegetables to be assessed [54].

7. Application of colorimeter

Colorimeter is the most common analytical technique used in biochemical estimation. It is used for quantitative estimation of protein, glucose, blood plasma, enzymes etc. Besides these biochemical estimations, colorimeter has a wide application of colorimetry in food industry. Over the last few years, the use of colorimeter has increased significantly to determine food color. Colorimeters are used for standardization of color, quality control inspection of various raw materials used such as meat, eggs, fishes, coffee, fruits and vegetables and color in every batch of final product specially in foods like jam, jelly, ketchup, beverages, baked goods, oil, sausages etc. Food products can be of different shape, size, granular or powder, solid and liquid of different viscosity. Thus, based on its optical characteristic food can be classified as opaque, translucent and transparent.

7.1 Tomatoes and its products

Tomato based products is thoroughly investigated on the basis of color than any other any other group of food color. Consumers usually associates freshness and maturity of raw tomatoes or the quality of its products by its redness, due to this reason the measurement of color is very important [55]. The grades of color to assess tomato juice, sauce, purees, ketchup or canned tomatoes was first done by MacGillivray by dating back during 1920s and 1930s on disc colorimeter [56, 57]. The measurement of the color was based on Munsell system. As described by Peng Wan, Roma and Pear varieties of tomatoes were taken having red, orange and green colors. About 150 images of the samples were taken. An image processing algorithm was developed using C++ and Matrox imaging library 9.0 (Matrox, Inc., Dorval, Canada).

Figure 11.
Setup capturing images of tomato [58].

Based on the images taken the color value was extracted and finally to develop a model on basis of which the maturity of the tomatoes could be classified [58]. **Figure 11** shows the setup of the image capturing process.

7.2 Fruits and its derivatives

Several studies have been conducted on the physicochemical changes occurring during the ripening of fruits and its harvesting. Color variables of unripe or ripe, pre-climatic or senescent variety and mangoes with post-harvest rapid ripening process could be co-related using CIE L*a*b* variables acquired by image analysis. This non-destructive method could adequately classify the ripening phases of mangoes [59].

Similarly, color content in the development of lemon skin such as chlorophyll, lutein and carotenoid content were measured with L*, a* and b* [60].

Image analysis data interpretation was done for raspberries at different maturation stages. The color, total antioxidant and polyphenolic compounds were evaluated using spectrophotometric methods, the result could determine the nutritional properties of this fruits [61].

The study on the change of carotenoid content (color degradation) of orange juice after and before pasteurization process and on storage at different temperature was evaluated by the CIEL*a*b*. This evaluation could detect the loss of carotenoid pigment and formation of brown color (browning) during pasteurization [62].

7.3 Meat and meat products

Meat color determination is a very crucial parameter of the quality assessment procedure as raw meat contains moisture, protein, vitamins and other nutrients which makes it susceptible to microbial spoilage. The color determines the freshness, tenderness and shelf life of the meat to be consumed. Freshly cut meat surfaces forms bright red color oxymyoglobin due to oxygenation from purple haem pigment myoglobin [63]. Oxymyoglobin further oxidizes to brownish color flesh metmyoglobin which is mostly rejected at the retailer and consumers. Measurement of the surface meat color was done using Munsell spinning discs in the year 1953 by Butler et al. [64], but with advancement in technology the process is done by using Minolta colorimeter, Video-meter lab (MSI) and machine vision.

7.4 Eggs

According to the study conducted by Milovanovic et al. [65] five species of eggs were taken: hen, duck, quail, goose and turkey. The samples were refrigerated for better separation of yolk and egg white [66]. At first the color of the egg shells was determined at seven random positions on each egg samples. After that the eggs were broken, the yolk and the albumin portion were separated and placed in a petri plate with white background. Color of all the three separated parts were measured colorimeter. The color reading was calculated as L*a*b, Whiteness Index (WI), Yellowing Index (YI) and total color difference (ΔE).

In case of hen's egg, the L* value detected was $L^* = 63.2$ which was closer to the duck's yolk $L^* = 62.0$ and both the yolks were brighter than quail's egg yolk where the L* value was $L^* = 51.6$. The intensity of red color was marked by the a* value where the hen's egg showed the highest intensity of $a^* = 4.0$, and highest b* value was $b^* = 42.4$. The egg yolk coloration depends on animal health, feed and capability to storage xanthophylls [67]. The feed of the bird determines the color of the egg yolk (yellow–red) which attributes to the amount of carotenoid and the antioxidant activity of the pigments [68]. The color of the yolk has a significant value from consumer point of view as the color is connected with its freshness and quality.

7.5 Color change due to thermal degradation

The changes in color due to heating can also be analyzed using colorimeter. The color change after heating cashew apple juice was quantified by a Spectrocolorimeter as L*, a* and b* values by CIE. Apart from CIE the Total color difference (ΔE*), Chrome (C*) and Hue (h) were also calculated [69]. The result showed reduction in the redness (a*) and yellowness (b*) after the juice was heat treated at 90°C. The simultaneous darkening of the juice indicated reduction in L* value, the total color change ΔE* increased as the time and temp increased. According to a study conducted by Lee and Coates [70], $\Delta E^* \approx 2$ represent color change but if the change is $\Delta E^* > 3$ then the product is unacceptable. Thus, consumer acceptability can be detected by colorimetric study.

7.6 Browning

Browning is a very important phenomenon in food processing such as baking, frying and drying effecting the flavor and color of the final product. Browning may result from enzymatic or non-enzymatic or Maillard browning occurring due to the oxidation of the phenolic compounds. Non-enzymatic browning is the chemical reaction between amino acid and reducing sugars by the impact of heat producing a golden brown to brown color with a desirable flavor. Formation of deep brown color while the food is charred leads to rejection of the product. Whereas, Enzymatic browning occurs by the action of enzyme polyphenol oxidase (PPO) which oxidizes the phenolic compounds in presence of air forming a pigment called melanin. Browning of fruits and vegetables may occur due to cutting, peeling or even due to any mechanical injury occurred. This browning reaction is undesirable and rejected by the consumers. It is beneficial and desirable only in case of coffee, cocoa, raisins, or tea when enzymatic browning contributes to the distinct color and flavor to the product.

The overall change is the brown color is denoted by Browning Index (BI) indicating the browning of food products due to presence of sugar. The CIE L*a*b* color space is used as the color model as the physical indicator of browning [48].

8. Conclusion

The link between human visual perception and food quality indicator was always considered to be the food color since time immemorial. Thus, color is the most important attribute influencing customer acceptability, food choices and purchase pattern. Measurement and analysis of color is an important factor in food industry to determine the use of food colorant, determine the quality and prescribed limit of artificial color in food, manage postharvest losses, understand the quality of food, determine the end point of baking, setting the time and temperature of a baking unit, determine the incubation or fermentation time of any food and so on. Colorimeter plays the major role in determining all these attributes of food quality. Color order systems that are widely adopted by the food industry are the HunterLab system, the CIEL*a*b* system, and the L*C*H* system. Recent advances in the field of image processing to determine machine vision is the future reality. Artificial Intelligence (AI) and machine vision are applied on both agriculture and food industries for generation of data and as a quality control parameter.

Acknowledgements

The authors would like to express their sincere gratitude to the President and the Director of the institution for their constant motivation and encouragement. Their expertise and insight were invaluable in shaping our research and to complete this chapter.

Conflict of interest

The authors declare that they have no known competing financial interests or personal relationships that could have appeared to influence the work reported in this paper.

Author details

Kakoli Dutta* and Rosalin Nath
Institute of Engineering and Management, Kolkata, India

*Address all correspondence to: kakoli.dutta@iem.edu.in

References

[1] Ohta N, Robertson A. Colorimetry: Fundamentals and Applications. England: John Wiley & Sons; 2006

[2] Giese J. Color measurement in foods. Food Technology Magazine, IFT. 2003; **57**(12):1-4

[3] Clydesdale FM, Ahmed EM. Colorimetry—methodology and applications. Critical Reviews in Food Science & Nutrition. 1978;**10**(3):243-301

[4] Iqbal A, Valous NA, Mendoza F, Sun DW, Allen. Classification of pre-sliced pork and Turkey ham qualities based on image colour and textural features and their relationships with consumer responses. Meat Science. 2010;**84**(3): 455-465

[5] Fathi M, Mohebbi M, Razavi SMA. Application of image analysis and artificial neural network to predict mass transfer kinetics and color changes of osmotically dehydrated Kiwifruit. Food and Bioprocess Technology. 2011;**4**:1357-1366

[6] Ghorpade VM, Deshpande SS, Salunkhe DK. Food colors. In: Food Additive Toxicology. Vol. 4. New York: Marcel Dekker Inc.; 1995

[7] Council, National Academy of Sciences. National Research, "Food Colors," Committee on Food Protection, Food and Nutrition Board, Division of Biology. 1971

[8] Carle R, Schweiggert R, editors. Handbook on Natural Pigments in Food and Beverages: Industrial Applications for Improving Food Color. UK: Woodhead Publishing; 2016

[9] Šivel M, Kráčmar S, Fišera M, Klejdus B, Kubáň V. Lutein content in marigold flower (*Tagetes erecta* L.) concentrates used for production of food supplements. Czech Journal of Food Sciences. 2014

[10] Leal F, de Clavijo CM. Annatto: Botany and horticulture. Horticultural reviews. 2011;**39**:389-419

[11] Bauernfeind JC. Natural food colors. In: Carotenoids as Colorants and Vitamin A Precursors. New York: Academic Press; 1981

[12] Fierens T, Van Holderbeke M, Cornelis C, Jacobs G, Sioen I, De Maeyer M, et al. Caramel colour and process contaminants in foods and beverages: Part II–Occurrence data and exposure assessment of 2-acetyl-4-(1, 2, 3, 4-tetrahydroxybutyl) imidazole (THI) and 4-methylimidazole (4-MEI) in Belgium. Food Chemistry. 2018;**255**:372-379

[13] Rodriguez-Amaya DB. Natural food pigments and colorants. In: Bioactive Molecules in Food. Cham: Springer; 2019. pp. 867-901

[14] Wu X, Prior RL. Systematic identification and characterization of anthocyanins by HPLC-ESI-MS/MS in common foods in the United States: Fruits and berries. Journal of Agricultural and Food Chemistry. 2005; **53**(7):2589-2599

[15] Wu X, Beecher GR, Holden JM, Haytowitz DB, Gebhardt SE, Prior RL. Concentrations of anthocyanins in common foods in the United States and estimation of normal consumption. Journal of Agricultural and Food Chemistry. 2006;**54**(11):4069-4075

[16] Strack D, Vogt T, Schliemann W. Recent advances in betalain research. Phytochemistry. 2003;**62**(3):247-269

[17] Biswas M, Dey S, Sen R. Betalains from *Amaranthus tricolor* L. Journal of Pharmacognosy and Phytochemistry. 2013;**1**(5):87-95

[18] Vaillant F, Perez A, Davila I, Dornier M, Reynes M. Colorant and antioxidant properties of red-purple pitahaya (*Hylocereus* sp.). Fruits. 2005;**60**(1):3-12

[19] Herbach KM, Rohe M, Stintzing FC, Carle R. Structural and chromatic stability of purple pitaya (Hylocereus polyrhizus [Weber] Britton & Rose) betacyanins as affected by the juice matrix and selected additives. Food Research International. 2006;**39**(6): 667-677

[20] Holden JM, Eldridge AL, Beecher GR, Buzzard IM, Bhagwat S, Davis CS, et al. Carotenoid content of US foods: An update of the database. Journal of Food Composition and Analysis. 1999;**12**(3): 169-196

[21] Furtado JD, Siles X, Campos H. Carotenoid concentrations in vegetables and fruits common to the Costa Rican diet. International Journal of Food Sciences and Nutrition. 2004;**55**(2): 101-113

[22] Reif C, Arrigoni E, Schärer H, Nyström L, Hurrell RF. Carotenoid database of commonly eaten Swiss vegetables and their estimated contribution to carotenoid intake. Journal of Food Composition and Analysis. 2013;**29**(1):64-72

[23] Xianquan S, Shi J, Kakuda Y, Yueming J. Stability of lycopene during food processing and storage. Journal of Medicinal Food. 2005;**8**(4): 413-422

[24] Lück E, Kuhnert P, Gölitz H, Grosse L, Jäger A, Otterstätter G, et al. Dictionary of Food Additives. Germany: Behr's Verlag; 1998

[25] Konig J. Food color additives of synthetic origin. In: Scotter MJ, editor. Color Additives for Food and Beverages. Woodhead Publishing; 2015. pp. 35-56

[26] Leulescu M, Rotaru A, Pălărie I, Moanță A, Cioateră N, Popescu M, et al. Tartrazine: Physical, thermal and biophysical properties of the most widely employed synthetic yellow food-colouring azo dye. Journal of Thermal Analysis and Calorimetry. 2018;**7**(8)

[27] EFSA, European Food Safety Authority. Guidelines on submission of a dossier for safety evaluation by the EFSA of active or intelligent substances present in active and intelligent materials and articles intended to come into contact with food. EFSA Journal. 2009:1208

[28] Wu L. Chapter 7—Analysis of food additives. In: Innovative Food Analysis. United Kingdom: Academic Press; 2021. pp. 157-180

[29] Scotter MJ. Methods for the determination of European Union-permitted added natural colours in foods: A review. Food Additives and Contaminants. 2011;**28**(5):527-596

[30] FAO. WHO Expert Committee on Food Additives and World Health Organization and Others. Evaluation of Certain Food Additives and Contaminants: Twenty-Seventh Report of the Joint FAO. Geneva: World Health Organization; 1983;(1)

[31] EFSA. Scientific opinion of the panel on biological hazards on the maintenance of the list of QPS microorganisms intentionally added to food or feed. EFSA Journal. 2009;7(12): 1-93

[32] Banerjee D, Chowdhary S, Chakraborty S, Bhattacharyya R. Recent advances in detection of food adulteration. Food Safety in the 21st Century. 2017:145-160

[33] Jain A, Pradhan BK, Mahapatra P, Ray SS, Chakravarty S, Pal K. Development of a low-cost food color monitoring system. Color Research & Application. 2021;**46**(2):430-445

[34] Blasco J, Aleixos N, Moltó E. Machine vision system for automatic quality grading of fruit. Biosystems Engineering. 2003;**85**(4):415-423

[35] Cubero S, Aleixos N, Moltó E, Gómez-Sanchis J, Blasco J. Advances in machine vision applications for automatic inspection and quality evaluation of fruits and vegetables. Food and Bioprocess Technology. 2011;**4**: 487-504

[36] Francis FJ, Markakis PC. Food colorants: anthocyanins. Critical Reviews in Food Science & Nutrition. 1989;**28**(4): 273-314

[37] Bartleson CJ. Memory colors of familiar objects. JOSA. 1960:73-77

[38] Berns RS. Billmeyer and Saltzman's Principles of Color Technology. England: John Wiley & Sons; 2019

[39] Wrolstad RE, Smith DE. Color analysis. Food analysis. 2017:545-555

[40] Quevedo R, Aguilera JM. Computer vision and stereoscopy for estimating firmness in the salmon (*Salmon salar*) fillets. Food and Bioprocess Technology. 2010;**3**:561-567

[41] n.d. CC BY-SA 3.0. Available from: https://commons.wikimedia.org/w/index.php?curid=1955750

[42] Gupte VC. Expressing colours numerically. In: Colour Measurement. United Kingdom: Woodhead Publishing; 2010. pp. 70-87

[43] Markovic I, Ilic J, Markovic D, Simonovic V, Kosanic N. Color Measurement Of Food Products Using CIE L*a*b* and RGB. Journal of Hygienic Engineering and Design. 2013;**4**(1):50-53

[44] SharkD. 2020. Available from: http://en.wikipedia.org/wiki/File:HSV_color_solid_cylinder.png.

[45] Randall D, Charlotte NC. Instruments for the measurement of color. In: Color Technology in the Textile Industry. North Carolina: AATCC; 1997. pp. 9-17

[46] Manamohana K, Bijur G, Malarout N, Singla B, Pavithra S, Naik N. Applications of colour models in the food processing. Palarch's Journal of Archaeology of Egypt/Egyptology. 2020; **17**(9):3496-3512

[47] EFSA. Recent developments in the risk assessment of chemicals in food and their potential impact on the safety assessment of substances used in food contact materials. European Food Safety Authority Journal. 2014:1-28

[48] Pathare PB, Opara UL, Al-Said FA. Colour measurement and analysis in fresh and processed foods: A review. Food Bioprocess Technology. 2013;**6**: 36-60

[49] Kumar V, Gill KD, Kumar V, Gill KD. Photometry: Colorimeter and spectrophotometer. In: Basic Concepts in Clinical Biochemistry: A Practical Guide. Switzerland: Springer; 2018. pp. 17-20

[50] Gallardo JC, Medina J. Interinstrumental agreement between spectrocolorimeters. In: 5th

Iberoamerican Meeting on Optics and 8th Latin American Meeting on Optics, Lasers, and Their Applications; Venezuela. SPIE; 2004;**5622**

[51] Kakani V, Nguyen VH, Kumar BP, Kim H, Pasupuleti VR. A critical review on computer vision and artificial intelligence in food industry. Journal of Agriculture and Food Research. 2020;**2**: 100033

[52] Sun D-W, editor. Computer Vision Technology for Food Quality Evaluation. UK: Academic Press; 2016

[53] Bhatt AK, Pant D. Automatic Apple Grading Model Development based on Back Propagation Neural Network and Machine Vision, and Its Performance Evaluation. AI & SOCIETY: Springer Nature; 2015. pp. 45-56

[54] Laykin S, Alchanatis V, Fallik E, Edan Y. Image-processing algorithms for tomato classification. ASABE. 2002;**45** (3):851-858

[55] Francis FJ, Clydesdale FM. Food Colorimetry: Theory and Applications. West Port, USA: AVI Publishing Co. Inc.; 1975

[56] MacGillivray JH. Tomato color as related to quality in the tomato canning industry. Agricultural Experiment Station: Purdue University; 1931; **350**(33)

[57] MacGillivray JH. Studies of tomato quality—IV: Variability in quality and food value of tomatoes. Proceedings of the Indiana Academy of Science. 1928; **38**:159-164

[58] Wan P, Toudeshki A, Tan H, Ehsani R. A methodology for fresh tomato maturity detection using computer vision. Computers and Electronics in Agriculture. 2018;**146**:43-50

[59] Vélez-Rivera N, Blasco J, Chanona-Pérez J, Calderón-Domínguez G, de Jesús P-FM, Arzate-Vázquez I, et al. Computer vision system applied to classification of "Manila" mangoes during ripening process. Food and Bioprocess Technology. 2014;**7**:1183-1194

[60] Conesa A, Manera FC, Brotons JM, Fernandez-Zapata JC, Simón I, Simón-Grao S, et al. Changes in the content of chlorophylls and carotenoids in the rind of Fino 49 lemons during maturation and their relationship with parameters from the CIELAB color space. Scientia Horticulturae. 2019;**243**:252-260

[61] Rodríguez-Pulido FJ, Gil-Vicente M, Gordillo B, Heredia FJ, González-Miret ML. Measurement of ripening of raspberries (*Rubus idaeus* L) by near infrared and colorimetric imaging techniques. Journal of Food Science and Technology. 2017;**54**(9):2797-2803

[62] Wibowo S, Vervoort L, Tomic J, Santiago JS, Lemmens L, Panozzo A, et al. Colour and carotenoid changes of pasteurised orange juice during storage. Food Chemistry. 2015;**171**:330-340

[63] MacDougall D. Colour in Food. Cambridge, England: Elsevier Science; 2002

[64] Butler OD, Bratzler LJ, Mallmann WL. The effect of bacteria on the color of prepackaged retail beef cuts. Food Technology. 1953;**7**(10):397-400

[65] Milovanovic B, Tomovic V, Djekic I, Solowiej BG, Lorenzo JM, Barba FJ, et al. Color assessment of the eggs using computer vision system and Minolta colorimeter. Journal of Food Measurement and Characterization. 2021;**15**:5097-5112

[66] Dvořák P, Doležalová J, Suchý P. Photocolorimetric determination of yolk

colour in relation to selected quality parameters of eggs. Journal of the Science of Food and Agriculture. 2009; **89**(11):1886-1889

[67] Baker R, Günther C. The role of carotenoids in consumer choice and the likely benefits from their inclusion into products for human consumption. Trends in Food Science & Technology. 2004;**15**(10):484-488

[68] An SY, Guo Y, Ma SD, Yuan JM, Liu GZ. Effects of different oil sources and vitamin E in breeder diet on egg quality, hatchability and development of the neonatal offspring. Journal of Animal Sciences. 2010;**23**(2):234-239

[69] Zepka LQ, Borsarelli CD, da Silva MAAP, Mercadante AZ. Thermal degradation kinetics of carotenoids in a cashew apple juice model and its impact on the system color. Journal of Agricultural and Food Chemistry. 2009; **57**(17):7841-7845

[70] Lee HS, Coates GA. LWT, effect of thermal pasteurization on Valencia orange juice color and pigments. Food Science and Technology. 2003;**36**(1): 153-156

Section 5

Applications of Colorimetric Analysis in Agriculture Enviornmental Medical and Pharmaceutical Industries

Chapter 6

Colorimetric Determination of Pyraclostrobin Fungicide Using P-Amino-sulphonic Acid Coupling Reagent in Agricultural Soil/Environmental Samples by Spectrophotometric Analysis

Chhaya Bhatt, Manish Kumar Rai and Joyce Rai

Abstract

A simple and sensitive method for the determination of pyraclostrobin, a widely used fungicide, is described here, which is based on diazotization and coupling with 4-aminosulfonic acid in alkaline medium. The reaction mechanism is based on the pre-equilibrium formation of amine and diazonium salt followed by a rate-limiting attack of the diazonium ion at an N-atom (N-coupling) to appear the corresponding red-colored azo complex. The λ_{max}, molar absorptivity and Sandell's sensitivity related to the UV-visible absorption spectrometry were found λ_{max} = 600 nm, 2.7×10^4 L mol^{-1} cm^{-1} and 1.01×10^{-5} µg cm^{-2}, respectively. Some of the important parameters like linearity range, limit of detection (LOD), limit of quantification (LOQ), correlation coefficient (R^2) and recovery% were calculated 3 to 12 µgmL^{-1}, 1.01µgmL^{-1}, 3.08 µgmL^{-1}, 0.984 and 93.5–99.3%, respectively, for the determination of organochlorine like pyraclostrobin using coupling reagent. The advantages of the present method are its simplicity, high selectivity and cost-effectiveness. In this article, the method has been validated by applying it to samples from different environmental conditions.

Keywords: spectrophotometry, pyraclostrobin, coupling reaction, recovery percentage, agricultural and environmental samples

1. Introduction

Pyraclostrobin (methyl N-[2-(1-(4-chlorophenyl)-1 h-pyrazol-3-yloxymethyl) phenyl](N-methoxy) Carbamate) (**Figure 1**) is a new and widely used fungicide of the strobilurin group [1]. Pyraclostrobin acts through inhibition of mitochondrial

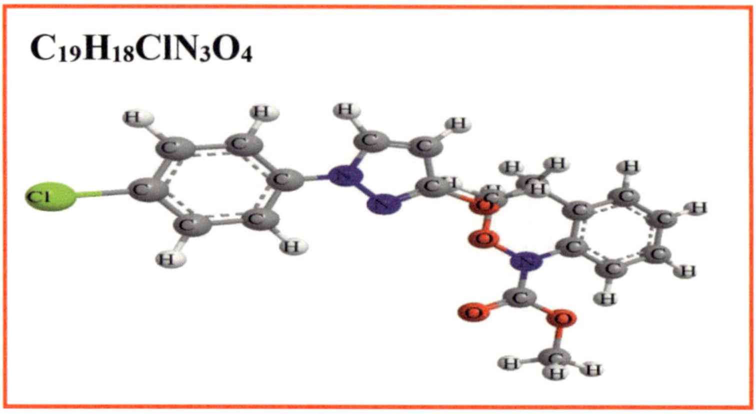

Figure 1.
Molecular formula and 3D structure of Pyraclostrobin.

respiration by blocking electron transfer within the respiratory chain that results incessation of fungal growth [2]. Despite their beneficial agricultural applications, agrochemicals in general. It can be extremely toxic to humans and animals and is very persistent in the environment [3]. Pyraclostrobin is a lipophilic fungicide that is harmful to aquatic creatures, particularly to fish. The distribution of pyraclostrobin residue in fish tissues under chronic toxicity has received significant attention in recent years, but little is known about its distribution in fish tissues under acute toxicity conditions [4]. Pyraclostrobin is a broad-spectrum fungicide that protects against powdery mildew, rust, web blotch, downy mildew and rice blast, which are all caused by oomycetes, ascomycetes, basidiomycetes, and asexual fungal species [5]. Currently, the pyraclostrobin is the most widely used fungicide group in the world, employed extensively in agricultural production [6]. The unreasonable usage of this fungicide has led to excess contaminants flowing into the aquatic environment by farming irrigation and surface runoff, which has caused fungicide accumulation in the soil and water ecosystem [7]. This contamination affects the aquatic organisms, food safety and human health in a negative way. As a result, pyraclostrobin determination research has become a prominent field of research [8].

Several techniques have been applied for the determination of pyraclostrobin, like Scanning electron microscope [9], rapid resolution liquid chromatography-tandem mass spectrometry [10], matrix solid-phase dispersion (MSPD) and gas chromatography, gas chromatography-mass spectrometry in selected ion monitoring mode (GC–MS, SIM) [11], QuEChERS and ultra-high-performance liquid chromatography-tandem mass spectrometry [12, 13]. However, these methods are sophisticated, expensive, time-consuming, and difficult to implement for the routine analysis of different matrices. This sheds light on the need for alternative simple and sensitive methods for pyraclostrobin analysis. Spectrophotometry is considered the most convenient analytical technique because of its inherent simplicity, low cost and wide availability in most quality control laboratories. Currently, a spectrophotometer method for determining pyraclostrobin in diverse environmental samples has been developed. In this article, we have proposed a spectrophotometer-based method for determining pyraclostrobin in diverse environmental samples.

2. Experimental section

2.1 Chemicals, reagents and solution preparations

Pyraclostrobin (purify 98%), sodium nitrite ($NaNO_2$) and 4-amino sulfonic acid were procured from Sigma-Aldrich (ACS reagent, ≥99%, MA, USA). Hydrochloric acid (HCl) was purchased from Hi-Media (AR reagent, 99% Mumbai, India). All the solutions were prepared in ultrapure water, and they were used throughout the studies. A stock standard solution of pyraclostrobin (1000 μgmL^{-1}) was prepared in an appropriate amount of analyte by dissolving in ultrapure water. Appropriate dilutions were made to obtain solutions containing pyraclostrobin in the concentration ranges from 3.0 to 12 μgmL^{-1}. About 1% $NaNO_2$ was prepared by dissolving 1.0 g of each targeted compound in 100 mL of ultrapure water. In addition, 1 N HCl solution was prepared by dissolving 8.3 mL of concentrated HCl in 100 mL of ultrapure water. A 1% solution of para-amino sulfonic acid (PAS) was prepared by taking 0.05 g of substance in 5.0 mL ethanol containing 10 mL of volumetric flask. To prepare a coupling reagent, 3 mL of HCl and 3 mL of $NaNO_2$ were mixed with 25 mL of solution containing 4-amino sulfonic acid reagent. All sample solutions were stored at 5°C until the analysis.

2.2 Sample collections and preparations

2.2.1 Liquid samples

Water samples were collected in bottles with special care avoiding air trapping, followed by sealing with Teflon-lined screw caps, kept in ice and transported to the laboratory before 24 h. The liquid sample is collected near the agricultural area. The liquid samples were stored at 3°C for spectrophotometric analysis.

2.2.2 Solid samples

The solid samples of potato, soybean, peanut, wheat grain, pea, banana, mango, papaya and environmental soil were collected using clean polyethylene bags and washed several times with distilled water. The samples were cut into small pieces and dried in hot-air oven for around 40 min. The samples were crushed with a pestle and homogenized. After this, 5.0 g of vegetables and soil samples were added to 50 mL water and kept overnight at room temperature. The next day, the sample was gently mixed with 50 mL ethanol and then centrifuged at 12000 rpm for 15 minutes. The supernatant part of the vegetable sample was filtered through a 0.45 mm Whatman filter paper and stored at 3°C for 24 hours. Finally, the prepared agricultural and environmental soil samples were used to quantitatively determine organochlorine (pyraclostrobin) using para-amino sulfonic acid (PAS) as a coupling reagent in spectrophotometric analysis.

2.3 Spectrophotometric determination of pyraclostrobin using coupling reagent

The schematic procedure for the determination of organochlorine using PAS as a coupling reagent in agricultural and environmental samples is depicted in **Figure 2**. For this, 1.0 mL of pyraclostrobin (4.0 $\mu g\ mL^{-1}$) was taken into 25 mL of clean and dry volumetric flask, and then 1.0 mL of a coupling reagent (PAS) was introduced

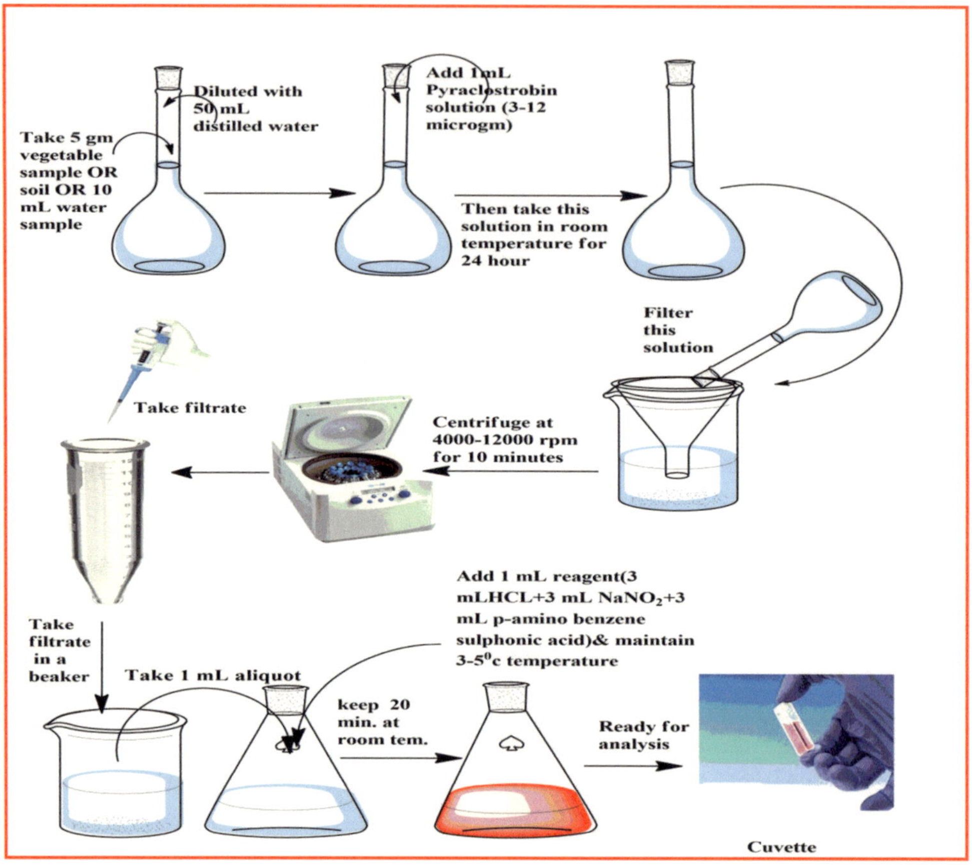

Figure 2.
The schematic representation for determination of pyraclostrobin.

into the sample under the 3–5°C temperatures. The pH of the sample solution was adjusted at 4.0 with the help of 1 N HCl and 1% NaOH solutions. The sample solution was turned to light red, indicating the formation of a color complex between pyraclostrobin and PAS as coupling reagents. The absorbance of the pyraclostrobin:PAS complex was measured in spectrophotometry at λ_{max} = 600 nm for the quantitative determination of pyraclostrobin. Subsequently, standard calibration curves of different concentrations of pyraclostrobin were prepared by performing different sets of experiments in triplicate sets of analyses by spectrophotometry. The blank sample was prepared under a similar condition. The λ_{max} = 600 nm for pyraclostrobin complex was selected for quantitative determination. The calibration curve was drawn between the different concentrations of pyraclostrobin (3.0 to12 µg mL^{-1}) and their respective absorbance or value of color intensity using linear least square (LLS) equation, i.e., y = mx + c. Regression analysis of Beer's law plots at their respective λ_{max} value revealed a good correlation.

2.3.1 Apparatus

The UV-vis absorption spectra were measured on a double beam spectrophotometer made of Cary-60 UV-vis spectrophotometer (Agilent Technologies) along with

99.99% accuracy of quartz glass. The absorption spectrum of the pyraclostrobin:PAS complex was recorded at 600 nm using a quartz cuvette with a path length of 1 cm. pH measurements were done using pH 700 EUTECH instrument. The centrifuge REMI R-4C was used at 12000 rpm for this coupling reaction. All glasswares were cleaned prior to use with an ultrasonic cleaning bath, PCI Analytics Pvt. Ltd., India, Model 3.5 L100H/DIC using mild detergents and rinsed with distilled water. The ultrapure water for solution preparation was obtained from a thermo fisher scientific Barnstead 234 Smart2Pure water system (conductivity 18.2 Ω 235^{-1}).

3. Results and discussion

3.1 Coupling reagent of pyraclostrobin

In the present method, the organic reagent dye (i.e., PAS) combines with a pesticide (pyraclostrobin) to form a pyraclostrobin:PAS complex by the coupled reaction through the diazonium ion (-N=N-). The color change of the sample solution from colorless to red indicated the formation of pyraclostrobin:PAS complex at a pH of 4.0 (acidic medium). The coupling reagent of pesticide (pyraclostrobin) showed a relatively narrow range of PAS reagent concentration that is suitable for quantitative analysis. Thereafter, UV-vis spectrophotometric with coupling reagent is a single-step extraction and pre-concentration process, recycled and the waste generated was easy to manage for the determination of pesticides like pyraclostrobin in different environmental and agricultural samples.

3.2 Spectral characteristic and quantitative screening of OCl (pyraclostrobin:PAS) complex

The UV-visible spectrophotometer was successfully applied for resolving complex mixtures for determination of pyraclostrobin in varieties of environmental and vegetable samples; recently, we have exploited the use of UV-vis spectrophotometry for the quantitative analysis of pesticide in environmental, vegetables and food samples based on the measurement of selected characteristic absorption bands of analyte complex with organic reagents in the UV-vis spectra [14]. In the present work, UV-visible spectrophotometer was used to acquire a suitable qualitative and quantitative analysis of pyraclostrobin as model compounds in an organic medium. For this, an aliquot of 1.0 mL from a test solution containing 3.0 to 12 μg mL^{-1} of pyraclostrobin in a 25 mL volumetric flask, and then 1.0 mL reagent such as PAS was introduced into the sample solution, resulting in the formation of pyraclostrobin:PAS complex. The ultrapure water was used to record the blank UV-vis absorption spectrum, which is used as a reference for the other spectral measurements. The absorption spectrum of the azo (pyraclostrobin:PAS) complex is depicted in **Figure 3**. The red color pyraclostrobin:PAS complex formed in the proposed reaction showed maximum absorbance at 600 nm. The color system obeyed Beer's law in the range of 3.0 to 12 μg mL^{-1}in the final solution volume of 25 mL. Here, the Beers–Lambert law explains the use of the terms absorbance, molar absorptivity, and Sandell's sensitivity relating to UV-visible absorption spectrometry, which were calculated and validated. The molar absorptivity and Sandell's sensitivity of red color pyraclostrobin:PAS complex were calculated to be 2.7×10^4 L mol^{-1} cm^{-1} and 1.0×10^{-5} μg cm^{-2}, respectively. The reproducibility of the method was assessed by carrying out six replicate analyses of a solution containing 4.0 μg mL^{-1} (**Table 1**).

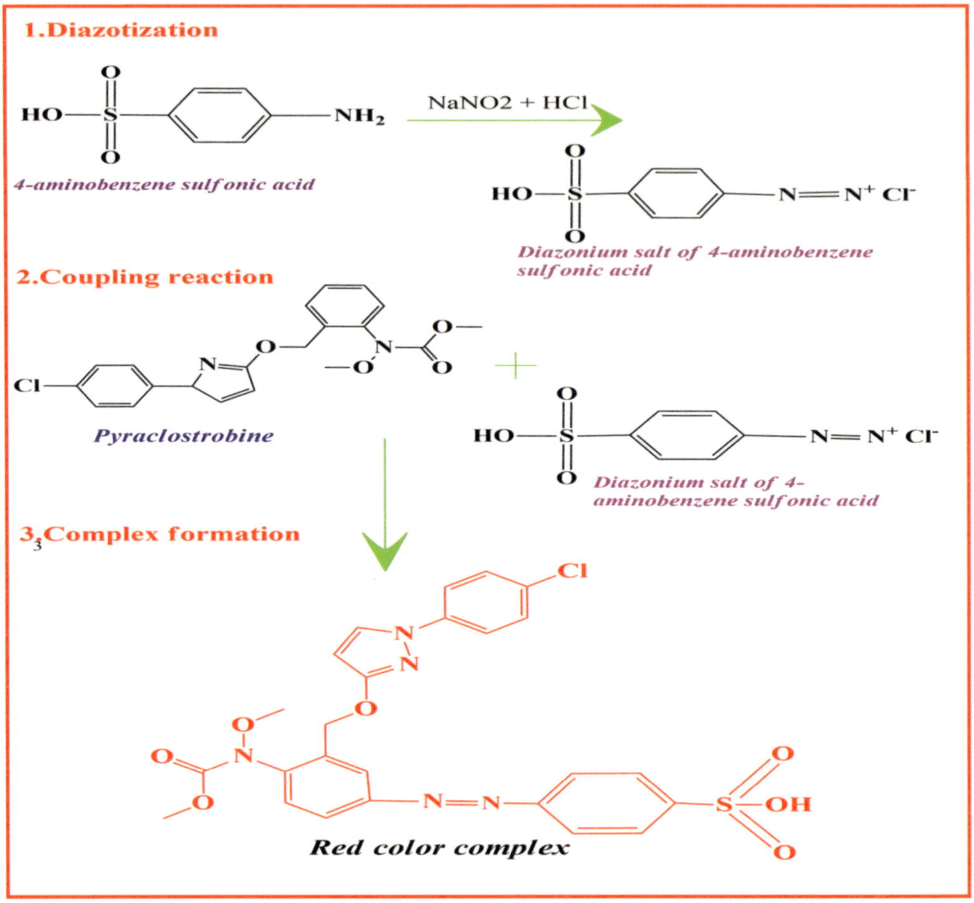

Figure 3.
The schematic mechanism of complex formation.

No. of days	Absorbance
1	0.709
2	0.715
3	0.719
4	0.707
5	0.722
6	0.727
7	0.730
Mean	0.718

Table 1.
Reproducibility of the method.

3.3 Mechanism for the determination of pyraclostrobin using PAS coupling reagent

In this method, para-amino sulfonic acid (PAS) is used as a new reagent for the formation of a color azo complex with target analytes, which is acts as a strong electrophilic in nature. PAS, a coupling reagent, reacts with OCl (pyraclostrobin) during the consumption of 4.0 mM of PAS dye to give a mixture of products. The remaining arene and azo groups reduced the color intensity of the reagent through disruption of the conjugation system in the PAS dye [15].

The schematic of the mechanism for selective detection of pyraclostrobin with PAS dye for coupling reaction is completed following step, and reactions are shown in **Figure 3**. The first step is diazotization process where the aromatic amine (p-amino sulfonic acid) is treated with nitrous acid or sodium nitrite, which is converted into nitrous acid in the presence of a strong acid like HCl at below 5°C temperature. This results in the loss of H_2O and the formation of new triple-bonded nitrogen atoms. This is followed by two proton transfers from nitrogen to oxygen accompanied by reorganization of the pi bonding framework such as forming N—N (pi) and breaking N—O (pi). This is based on the greater stability of aromatic diazonium salts than aliphatic diazonium salts because the electron-rich benzene ring stabilizes the N≡N group or delocalized system [16]. If the temperature rises above 5°C, the benzene diazonium chloride decomposes to form phenol and nitrogen gas is given off (**Figure 3**). The second step is the coupling reaction, where a benzene diazonium chloride reacts with a benzene ring having pyraclostrobin as a pesticide to produced red color of azo compound. The mechanism involves an initial attack of a coupling agent (phenols or anilines) on an electrophilic diazonium ion, followed by the loss of a proton. This process is known as the coupling reaction (**Figure 3**). The reason for PAS reagent couples with pyraclostrobin in an acidic medium and gives the red color of pyraclostrobin:PAS azo complex is following. The diazonium salts act as an electrophilic nature and attack the nucleophilic site during the reaction. A colored precipitate of azo compound is formed immediately on reaction of diazonium salt with amines or phenols. Azo dyes are compounds that contain two aromatic fragments connected by an N=N double bond. Azo compounds are stable, so the dyes do not fade.

UV-vis spectrophotometry was used to reduce interference and avoid overlapped absorption bands during the determination of pyraclostrobin in an aqueous medium in the visible regions. This theory is based on energy absorbed, which causes changes in electronic energy when the electrons are promoted from a ground state to an excited state [17]. The chromophoric functional groups are responsible for giving color due to the electrons excitation of the outer bonding electrons or lone pairs to these dyes. The auxochromes are the functional groups that are an integral part of PAS reagents or dyes and are also responsible for intensifying the color and giving UV-vis spectra at 600 nm (λ_{max}). Thus, the use of the color pyraclostrobin:PAS complex was successfully demonstrated for analysis of Ocl, for example, pyraclostrobin in UV-vis spectrophotometry.

3.4 Optimization and analytical parameters

The coupling reagent of pyraclostrobin from sample solution was monitored by considering the absorption band obtained at 600 (λ_{max}) in UV-vis spectrophotometry. The optimum conditions for analytical parameters such as reagent concentrations, volume of reagent, volume of sample, effect of pH, extraction time and stirring rate were investigated for obtaining the efficient extraction of pesticides like pyraclostrobin from environmental and agricultural samples. The influences of each of the following variables on the reaction were tested.

3.4.1 Effect of PAS reagent concentration and volume

The effect of concentration and volume of PAS reagent were investigated for optimum coupling reaction as well as formation of stable pyraclostrobin:PAS complex by varying the concentration of reagent from 1.0 to 5.5 mM and volume of reagent from 0.2 to 2.0 mL. The absorbance of reaction solution increases with the reagent concentration, and the highest absorption intensity was attained at 5.5 mM. Similarly, a 1.0 mL optimum volume of PAS reagent is used for the production of maximum and reproducible color complex. Therefore, the 5.5 mM reagent concentration with 1.0 mL volume was used throughout the work (**Figures 4** and **5**).

3.4.2 Influence of variable pH as an extracting medium

The reactivity of pyraclostrobin was investigated by the effect of different pH values in the range of 2.0 to 6.5 during the extraction and pre-concentration of pyraclostrobin from real samples (**Figure 6**). The color intensity of the sample solution was changed with the increases in pH up to 4.0 to 6.0 and beyond the suddenly decreased due to the repulsive force of ions. In addition, the maximum absorption band was obtained when maintaining the pH 4.0 at 3°C temperature, indicating the formation of stable pyraclostrobin:PAS complex at acidic conditions. Therefore, the concentration at pH 4.0 was selected for the quantitative analysis through UV-vis spectrophotometry.

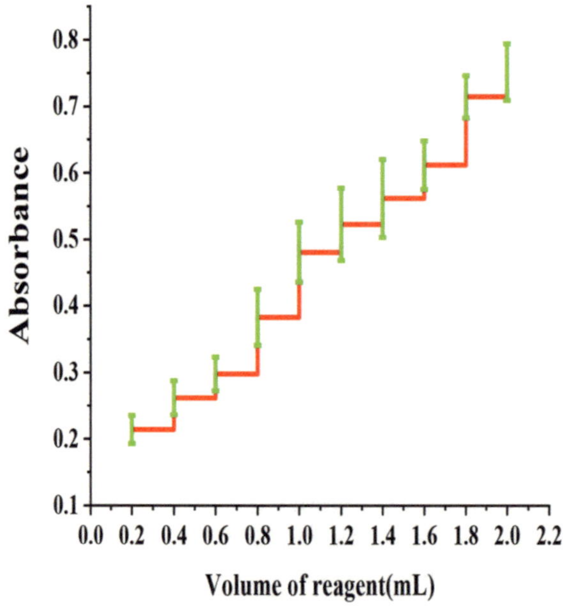

Figure 4.
Effect of volume of reagent.

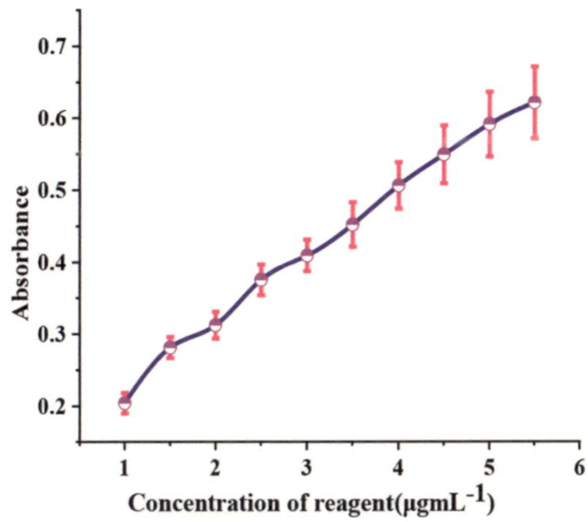

Figure 5.
Effect of concentration of reagent.

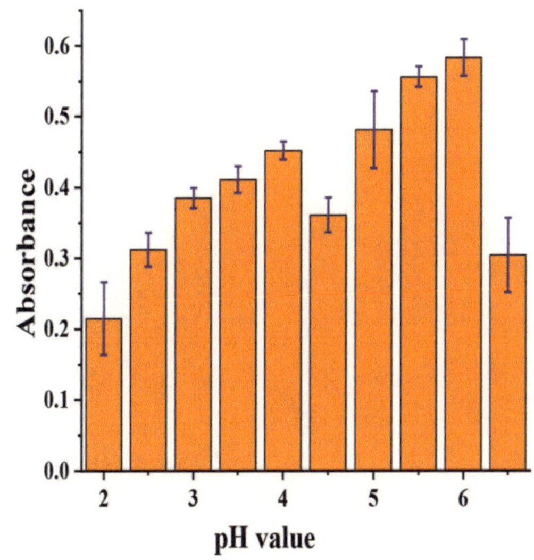

Figure 6.
Effect of pH.

3.5 Effect of interfering ions and cross-contaminants

Under the optimized experimental conditions, we investigated the interference study that was carried out for selective determination of pyraclostrobin as OCl in sample solution using PAS as coupling reagent. Selectivity describes whether methodology could discriminate interference of similar groups of compounds in the presence of analytes, which is determined by relative absorbance value by spectrophotometric method [18]. To assess the validity of the proposed method, the effects of various common foreign species and other pesticides were added to the standard solution containing 4.0 μg mL^{-1} of pyraclostrobin prior to the coupling reaction and analyzed by the spectrophotometric method at 600 wavelengths. The tolerance limits of different foreign species ions, along with other classes of pesticides, are

Parameters	Values for the reaction
λ_{max}, nm	600
Beer's law limit, μg mL^{-1}	3–12 μg/10 mL
Detection limits, μg mL^{-1}	0.077
Quantification limits, μg mL^{-1}	0.235
Molar absorptivity, L mol^{-1} cm^{-1}	2.7×10^4
Sandell's sensitivity, μg cm^{-2}	1.0×10^{-5}
Regression equation, $y = bx + a$	$y = 0.026x + 3.089$
Intercept (a)	3.089
Slope (b)	0.026
Standard deviation	0.008
Relative standard deviation (%)	1.11
Correlation coefficient (r^2)	0.984

Table 2.
Optical characteristics and statistical data for the reaction.

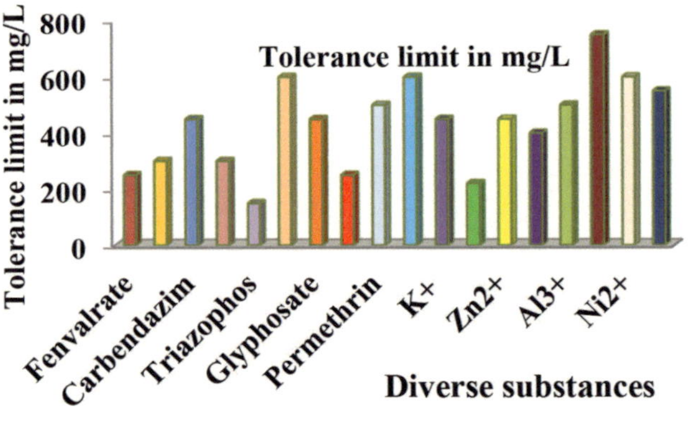

Figure 7.
Effect of various pollutants and pesticides.

given in **Table 2** and **Figure 7**. The peak intensity or maximum λ_{max} of pyraclostrobin remained unchanged in the presence of tested organic and inorganic chemical species at the optimized conditions of the proposed method. In addition, there was no color change with the PAS coupling reagent, or no coupled reaction was found in the presence of other classes of pesticides under the optimized conditions, at the same time as only pyraclostrobin displayed the color change from colorless to red color during the coupling reaction (**Figure 2**). Therefore, the spectrophotometric method was found free from interference of diverse substances and other pesticides commonly associated with determining OCl like pyraclostrobin from environmental and agricultural samples.

3.6 Analytical validation (determination) of pyraclostrobin using spectrophotometry

Some of the essence parameters such as linearity range, molar adsorptive and Sandell's sensitivity, accuracy and precision, robustness and ruggedness, limit of detection (LOD), limit of quantification (LOQ), correlation coefficient (R^2) and correlation estimation (R) for the determination of pyraclostrobin were estimated to determine the plausibility of using coupling reagent.

3.6.1 Linearity range, molar adsorptive and Sandell's sensitivity

The linearity range was obtained by constructing the calibration curve using the absorbance value versus different concentrations of pyraclostrobin, shown in **Figure 8**. The statistical parameters were given in the regression equation (y = mx + c) calculated from the calibration graphs, along with the standard deviation (SD) of the slope (m) and the intercept (c) on the ordinate and the SD residuals (SDy/x). The linearity of calibration graphs was proved by high values of the correlation coefficient (r) and small values of the y-intercepts of the regression equations. The absorption spectrum of the red color pyraclostrobin:PAS complex is shown in **Figure 9**, with maximum absorption at 600 nm. The apparent molar absorptivity of the resulting colored complex and RSD % of response factors for the spectrophotometric method were also calculated and validated (**Table 2**). Beer's law was obeyed, and the linearity graph is shown in **Figure 9** in the concentration range of 3.0 to 12 µg/1.0 mL of Pyraclostrobin solution. We have also calculated absorptivity and Sandell's sensitivity (**Table 3**).

3.6.2 Accuracy and precision

The accuracy of the method was determined by calculating the recovery percentage (%) of pyraclostrobin and by comparing the results with standard reference method (ICH Q2 R1 standard guideline) [19, 20]. The recovery % for pyraclostrobin method was calculated by spiking definite concentrations in real environmental and agricultural samples. A good recovery % obtained with coupling reagent

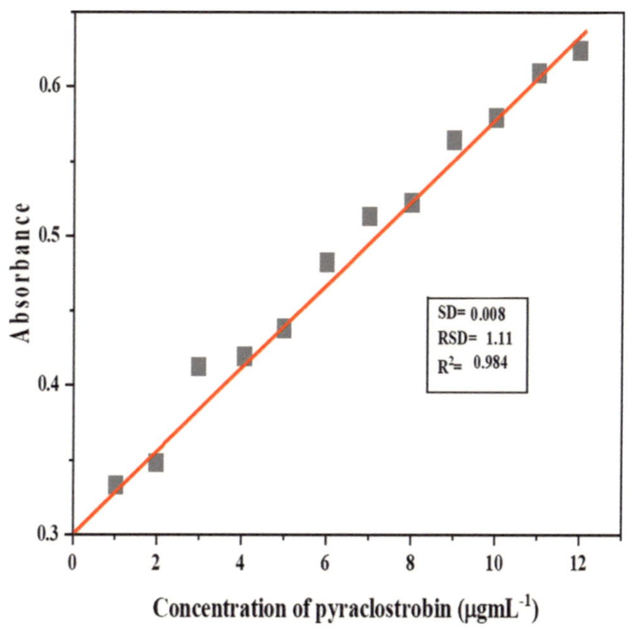

Figure 8.
Calibration curve for the different concentrations of pyraclostrobin (3–12 µg mL^{-1}).

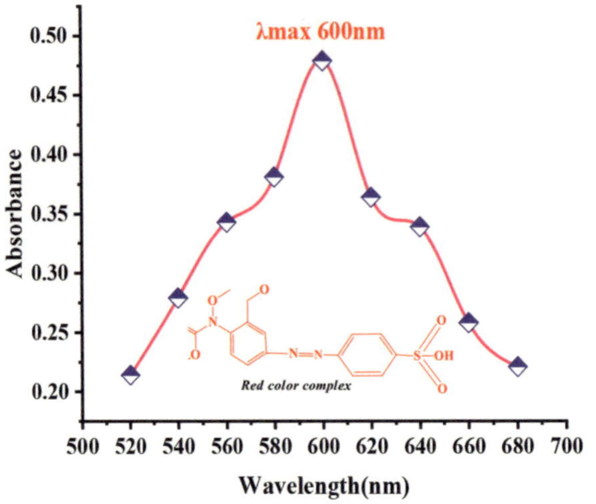

Figure 9.
UV-visible absorption spectrum of red-colored dye complex.

Foreign species	Tolerance limit in mg/L	Foreign ions	Tolerance limit in mg/L
Fenvalrate	250	Na^+	600
Ethion	300	K^+	450
Carbendazim	450	Ca^{2+}	220
Acetampride	300	Zn^{2+}	450
Triazophos	150	Ba^{2+}	400
Kitazin	600	Al^{3+}	500
Glyphosate	450	SO_4^{2-}	750
Metribuzin	250	Ni^{2+}	600
Permethrin	500	Cl^-	550

Table 3.
Effect of foreign species and ions.

coupled with spectrophotometry was calculated in the range of 84.75–108.5% for pyraclostrobin in agricultural and environmental samples, as displayed in **Table 4**. The precision is another parameter that should be validated to determine the reproducibility of the results (ICH Q2 R1 standard guideline). The precision of the proposed method was ascertained by actual determination of fixed concentration of the pesticide within Beer's range and finding the absorbance. The different levels of pesticide concentration (six times) are prepared three different times in a day and studied for intra-day variation and three different days to study inter-day variation. The percentage relative standard deviation (% RSD) of the predicted concentration from the regression equation is taken as precision. Accordingly, the RSD % was found to be 1.9% for pyraclostrobin. These results show the precision, accuracy and selectivity for the determination of OCL-like pyraclostrobin in agricultural and environmental samples.

Sample	Pyraclostrobin in original found (X)	Pyraclostrobin added in µg (Y)	Total Found (Z)	Difference (µg) (Z-X)	Recovery (%) (Z-X)/Yx100	RSD
Water**	0.26	10	9.72	9.45	94.5	0.395
Soil***	0.21	10	9.78	9.56	95.6	0.170
Potato***	0.52	10	10.38	9.86	98.6	0.396
Soyabean***	0.31	10	10.24	9.93	99.3	0.408
Peanut***	0.38	10	10.20	9.82	98.2	0.764
Wheatgrain***	0.24	10	10.15	9.91	99.1	0.510
Pea***	0.25	10	9.83	9.58	95.8	0.396
Banana***	0.43	10	10.30	9.87	98.7	0.663
Mango***	0.27	10	9.72	9.44	94.4	0.172
Papaya***	0.32	10	9.67	9.35	93.5	0.174

**Mean of three replicate Analysis Amount of sample 50 mL.*
***Amount of sample 5 gm.*
****Amount of sample 5 gm.*

Table 4.
Recovery of pyraclostrobin in environmental samples.

3.6.3 Reproducibility

The solution containing 4.0 µg mL^{-1} of pyraclostrobin was evaluated for reproducibility by performing three replicate analyses for 7 days, and then standard deviation (±SD) and relative standard deviation (RSD%) were calculated at ±0.008 and 1.11%, respectively (**Table 3**). For this wavelength, negligible absorbance was observed for blank reagents. We did not observe any color dye azo formation in the presence of other classes of pesticides. In the study area, maximum absorption was obtained at 600 nm. The main reasons are less spectral interference and high absorption of pyraclostrobin in an aqueous medium. The measurement repeatability expresses the closeness of the results obtained with the same sample using the reagent coupled with spectrophotometric analysis in a short period of time to give the smallest possible variation in all experimental results.

3.6.4 Robustness and ruggedness

The robustness and ruggedness were evaluated for the coupling reagent coupled with spectrophotometric determination of pyraclostrobin in agricultural and environmental samples. For evaluation of the method's robustness, the different parameters like pH, reagent concentration, wavelength range and time remain unaffected by small, deliberate variations. Method ruggedness was expressed as RSD % by six analysts using both spectrophotometric methods on different days. The results showed no statistical differences between instruments, suggesting that the developed methods were robust and rugged (**Table 3**).

3.6.5 LOD and LOQ

LOD is defined as a minimum quantity of substance that the instrument can respond to with a signal-to-noise ratio of three (S/N = 3) (or $3\sigma/k$); LOQ is the minimum quantity of substance when S/N = 10 (or10σ/k) with measured precision during routine laboratory operating conditions [21]. Where σ is the standard deviation of replicate analysis values under the same conditions as for the sample analysis in the absence of the analyte, and k is the sensitivity, namely the slope of the calibration graph. The value of LOD and LOQ in the present work was calculated to be 1.01 and 3.08 μgmL^{-1}, respectively (**Table 3**).

3.7 Applications to agricultural and environmental samples

The recommended method was successfully applied to the determination of OCl-like pyraclostrobin in various agricultural and environmental samples from rural residential sites of Raipur city. The reason for choosing this region is that a large number of crops are produced frequently. The details for the sample preparation procedure of agricultural and environmental samples are discussed in the experimental section and **Figure 2**. Finally, the prepared vegetables, soil, and environmental water samples were used for the quantitative determination of OCl (pyraclostrobin) using coupling reagent in the spectrophotometric method. The absorbance of the red color of the sample containing pyraclostrobin:PAS complex was measured in spectrophotometry at λ_{max} = 600 nm for quantitative determination of pyraclostrobin in vegetable and environmental samples at a very trace level under optimum conditions, which is displayed in **Figure 9** followed by the spectrophotometric method. Ten agricultural and environmental samples were tested, which were obtained from rural farm sites of Raipur city, Chhattisgarh. Five agricultural and two environmental samples were found to be positive toward the presence of pyraclostrobin in significant concentrations (**Table 4**). The concentrations of pyraclostrobin were determined in pure real samples, and in the same samples after spiking, the standard concentration is 4.0 μgmL^{-1}. Blank samples, those containing undetectable quantities of pesticide, were analyzed to evaluate the selectivity of the proposed coupling reagent coupled with spectrophotometric method. In addition, the blank samples did not show

Methods	Linear range, μgmL^{-1}	LOD, μgmL^{-1}	LOQ μgmL^{-1}	correlation coefficient (R^2)	Samples	Ref.
Solid-phase micro extraction (SPME) GC-MS	1.00–100.0	0.30	1.00	0.993	Environment	[22]
Ultra-performance liquid chromatography tandem mass spectrometry UPLC-MS/MS	0.0001–0.1	0.70	2.23	0.999	aquatic ecological	[4]
LC-MS/MS	0.2–200	0.06	0.2	0.998	Environment	[23]
UV-visible spectrophotometer	3–12	0.077	0.235	0.984	Environment and agricultural	Present method

Table 5.
The comparison of present UV-Visible method with other reported methods for the determination of pyraclostrobin in varieties of the samples.

maximum wavelength (λ_{max}) at 600 nm, indicating the absence of the analytes. The concentration of pyraclostrobin was obtained using the linear regression equation model (**Figure 8**).

Table 5 provides the analytical data for determination of pyraclostrobin using coupling reagents from agricultural and environmental samples in spectrophotometry and compared with the results of other reported methods in terms of linearity range, LOD and sample matrices. In the present work, coupling reagent was performed without using toxic solvents in spectrophotometry for the determination of pyraclostrobin in different agricultural and environmental samples at very microgram (μg) levels as compared to other instrumental methods such as solid-phase microextraction (SPME) GC-MS, ultra-performance liquid chromatography-tandem mass spectrometry UPLC-MS/MS and HPLC-UV (**Table 5**). The results indicate that a many-fold enhancement in the recovery % and extraction efficiency will be acquired using PAS coupling reagent-assisted microextraction of pyraclostrobin as compared to the other methods. The present method based on PAS:pyraclostrobin complex is found to be simple, eco-friendly, sustainable, selective, sensitive and cost-effective as compared to different spectroscopic and chromatographic methods [22, 24–26].

4. Conclusion

The present investigations were carried out with a view to examining the potential of para-amino sulfonic acid (PAS) as a coupling reagent for OCl-like pyraclostrobin present in various agricultural and environmental samples and its subsequent determination by coupling reagent coupled with spectrophotometric method. The pyraclostrobin:PAS complex is very sensitive for the determination of pyraclostrobin at the lowest concentration, and it can be easily detected by the naked eye due to the prominent color variation of the sample solution with and without the addition of analytes. The routine analysis of OCl is normally performed using sophisticated instruments such as GC-MS, LC/ESI-MS, HPLC, RS-TLC and UHPLC/QqTOF-MS, which are generally large in size, require rigorous training to operate, require large quantities of chemicals and reagent, and involve time-consuming sample preparation processes. Our strategic endeavor was initiated with a coupling reagent followed by the analysis of pyraclostrobin using a convenient and simple spectrophotometric method without any requirement for specific cleanup and sample preparation steps. The current approach is simple, cost-effective and does not require specific sample preparation for better extraction, leading to enhanced extraction and recovery % in UV-visible regions during the analysis. The method has shown good sensitivity, reliability and ease of preparation of the reagent compared with other existing extractive spectrophotometric determination methods.

Author details

Chhaya Bhatt[1*], Manish Kumar Rai[1*] and Joyce Rai[2]

1 School of Studies in Chemistry, Pt. Ravishankar Shukla University, Raipur, Chhattisgarh, India

2 Chhattisgarh Council of Science and Technology, Raipur, Chhattisgarh, India

*Address all correspondence to: chhaya05bhatt@gmail.com and mkjkchem@gmail.com

References

[1] Reddy SN, Gupta S, Gajbhiye VT. Effect of moisture, organic matter, microbial population and fortification level on dissipation of pyraclostrobin in soils. Bulletin of Environmental Contamination and Toxicology. 2013;**91**:356-361. DOI: 10.1007/s00128-013-1045-0

[2] Jimenez-Lopez J, Llorent-Martínez EJ, Ortega-Barrales P, Ruiz-Medina A. Sensitive photochemically induced fluorescence sensor for the determination of nitenpyram and pyraclostrobin in grapes and wines. Food Analytical Methods. 2019;**12**:1152-1159. DOI: 10.1007/s12161-019-01451-5

[3] Lopes FM, Batista KA, Batista GLA, Mitidieri S, Bataus LM, Fernandes KF. Biodegradation of epoxyconazole and piraclostrobin fungicides by *Klebsiella* sp. from soil. World Journal of Microbiology and Biotechnology. 2010;**26**:1155-1161. DOI: 10.1007/s11274-009-0283-0

[4] Li H, Yang S, Li T, Li X, Huanga X, Gao Y, et al. Determination of pyraclostrobin dynamic residual distribution in tilapia tissues by UPLC-MS/MS under acute toxicity conditions. Ecotoxicology and Environmental Safety. 2020;**206**:111182. DOI: 10.1016/j.ecoenv.2020.111182

[5] Gao YY, Li XX, He LF, Li BX, Liu F. Effect of application rate and timing on residual efficacy of pyraclostrobin in the control of pepper anthracnose. Plant Disease. 2020;**104**(610-1001):958-966. DOI: 10.1094/PDIS-03-19-0435-R

[6] Gao YY, He LF, Li BX, Mu W, Lin J, Liu F. Sensitivity of colletotrichum acutatum to six fungicides and reduction in incidence and severity of chili anthracnose using pyraclostrobin. Australasian Plant Pathology. 2017;**46**:521-528. DOI: 10.1007/s13313-017-0518-8

[7] Gao YY, Li XX, He LF, Li BX, Mu W, Liu F. Role of adjuvants in the management of anthracnose-change in the crystal morphology and wetting properties of fungicides. Journal of Agricultural and Food Chemistry. 2019;**67**:9232-9240. DOI: 10.1021/acs.jafc.9b02147

[8] Mercader JV, Suárez-Pantaleón C, Agulló C, Abad-Somovilla A, Abad-Fuentes A. Production and characterization of monoclonal antibodies specific to the strobilurin pesticide pyraclostrobin. Journal of Agricultural and Food Chemistry. 2008;**56**(17):7682-7690. DOI: 10.1021/jf801340u

[9] Yan X, Wang Y, Liu H, Li R, Qian C. Synthesis and characterization of melamine-formaldehyde microcapsules containing pyraclostrobin by *in situ* polymerization. Polymer Science Series B. 2018;**60**(6):798-805. DOI: 10.1080/10601325.2019.1596746

[10] Fan X, Zhao S, Chen X, Hu J. Simultaneous determination of pyraclostrobin, prochloraz, and its metabolite in apple and soil via RRLC-MS/MS. Food Analysis Methods. 2018;**11**:1312-1320. DOI: 10.1007/s12161-017-1065-1

[11] Lagunas-Allue L, Sanz-Asensio J, Martinez-Soria MT. Response surface optimization for determination of pesticide residues in grapes using MSPD and GCMS: Assessment of global uncertainty. Analytical and Bioanalytical Chemistry. 2010;**398**:1509-1523. DOI: 10.1007/s00216-010-4046-4

[12] Kemmerich M, Rizzetti TM, Martins ML, Prestes OD, Adaime MB, Zanella R. Optimization by central composite design of a modified QuEChERS method for extraction of pesticide multiresidue in sweet pepper and analysis by ultra-high-performance liquid chromatography–tandem mass spectrometry. Food and Analytical Methods. 2015;**8**:728-739. DOI: 10.1016/j.foodchem.2015.09.010

[13] Barbosa PGA, Itana F, Martins CC, Lima LK, Milhome MAL, Cavalcante RM, et al. Statistical analysis for quality adjustment of the analytical curve for determination of pesticide multiresidue in pineapple samples. Food Analytical Methods. 2018;**11**:466-478. DOI: 10.1007/s12161-017-1017-9

[14] Sahu B, Kurrey R, Deb MK, Shrivas K, Karbhal I, Khalkho BR. A simple and cost-effective paper-based and colorimetric dual-mode detection of arsenic(III) and lead(II) based on glucose-functionalized gold nanoparticles. RSC Advances. 2021;**11**(34):20769-20780. DOI: 10.1039/D1RA02929K

[15] Sharma H, Saha A, Mishra AK, Rai MK, Deb MK. Diazotized reagent for spectrophotometric determination of glyphosate pesticide in environmental and agricultural samples. Journal of the Indian Chemical Society. 2022;**99**(7):100483, ISSN 0019-4522,. DOI: 10.1016/j.jics.2022.100483

[16] Kumar V, Kaur S, Singh S, Upadhyay N. Unexpected formation of N'-phenyl-thiophosphorohydrazidic acid O,S-dimethyl ester from acephate: Chemical, biotechnical and computational study. 3 Biotech. 2015;**6**(1):1. DOI: 10.1007/s13205-015-0313-6

[17] Sharma H, Mishra AK, Rai MK. Spectrophotometric determination of acephate with 4-aminoazobenzene and its applications. GIS Science Journal. 2022;**9**:2, ISSN NO: 1869-9391. DOI: 20.18001.GSJ.2022.V9I2.22.38725

[18] Goswami J, Banjare MK, Banjare RK, Rai JK, Rai MK. Extraction of acephate pesticide in environmental and agricultural samples by spectrophotometric method. Journal of the Indian Chemical Society. 2021;**98**(9):100138. DOI: 10.1016/j.jics.2021.100138

[19] Nardelli V, Oro D, Palermo C, Centonze D. Multi-residue method for the determination of organochlorine pesticides in fish feed based on a cleanup approach followed by gas chromatography-triple quadrupole tandem mass spectrometry. Journal of Chromatography. A. 2010;**1217**:4996-5003. DOI: 10.1016/j.chroma.2010.05.052

[20] Von Holst C, Müller A, Björklund E, Anklam E. In-house validation of a simplified method for the determination of PCBs in food and feeding stuffs. European Food Research and Technology. 2001;**213**(2):154-160. DOI: 10.1007/s002170100326

[21] Janghel EK, Rai JK, Rai MK, Gupta VK. A new and highly sensitive spectrophotometric determination of monocrotophos in environmental and biological samples. Journal of Chinese Chemical Society (China). 2006;**53**:343-347

[22] Mamta Nirmal MK, Rai VP, Rai JK. A new spectrophotometric method for the determination of isoprothiolane in environmental sample. International Journal of Science and Research. 2014;**3**(5):1820-1822

[23] Gao YY, He LF, Li BX, et al. Sensitivity of Colletotrichum acutatum to six fungicides and reduction

in incidence and severity of chili anthracnose using pyraclostrobin. Australasian Plant Pathol. 2017;**46**:521-552. DOI: 10.1007/s13313-017-0518-8

[24] Janghel EK, Rai JK, Rai MK, Gupta VK. A new solid sorbent system for rapid monitoring of nicotine. Journal of Indian Chemical Society. 2005;**82**(11):1032-1034

[25] Janghel EK, Rai JK, Khan S, Rai MK, Gupta VK. TLC-spectrophometric separation and trace determination of monocrotophos and dichlorvos in environmental and biological samples. Journal of Environmental Health, India. 2007;**49**(2):133-141

[26] Vindhiya Patel MK, Rai PM, Rai JK. Sensitive spectrophotometric determination of chloropyriphos in different environmental and biological sample. International Journal of Science and Research. 2014;**3**(7):1545-1550. Available from: https://www.ijsr.net/get_abstract.php?paper_id=20141126

Section 6

Application of Colorimetric Analysis in Textile and Apparel Industry

Chapter 7

Study on the Compatibility of a Binary Mixture of Babul Bark and Other Natural Dyes Used on Gallnut Pre-Mordanted Cotton Fabric for Developing Compound Shades

Deepali Singhee, Yamini Dhanania and Ashis Kumar Samanta

Abstract

Cotton fabric that has been pre-mordanted with gallnut extract was dyed using a binary mixture of aqueous extracts of the natural dyes such as babul bark with bixa (annatto) seeds, *Rheum emodi* (rhubarb) roots and pomegranate rind extracts to obtain compound shades. The compatibility of each binary pair of natural dyes was assessed by (a) a conventional process of analysis through progressive development of shades by changing time and temperature in one set, and variation in the total concentration of dyes in the mixture in the second set with 50:50 dye proportion and (b) by a newer approach of analysing the relative compatibility rating (RCR) of each pair of dyes by varying proportion of both the dyes (75:25, 50:50, 25:75) in the mixture. The study indicates that all three pairs of binary mixtures (babul bark with annatto seeds, babul bark with pomegranate rind and babul bark with Himalayan rhubarb roots) are found to vary in their respective degree of compatibilities and are differently compatible (moderated, fair and average) as determined by the newer RCR method. Interestingly the degree of compatibility found by the conventional method nearly matches with the newer method showing the efficacy of the RCR method/technique.

Keywords: bixa orellana (annatto), babul bark, binary mixture of natural dyes, compatibility of natural dyes, cotton, pomegranate, relative compatibility rating, *Rheum emodi* (Himalayan rhubarb)

1. Introduction

Despite a number of drawbacks, natural colourants have been used to dye textiles since ancient times. In recent years, the use of natural dyes has experienced a significant resurgence due to the chemical hazards and harmful effects of some synthetic dyes [1]. Natural dyes offer a restricted range of shades; thus, textile dyers must create a wide range of colours to meet customer demands, colour forecasts, and fashion trends. Development of compound shades using natural dyes involves sequential dyeing in multiple baths, which is invariably time-consuming, expensive and associated with increased fabric damage. Therefore, there is a need for an alternate method that allows dyers to blend natural dyes in a single bath to create wider spectrum of colour shades.

Researchers have investigated the compatibility of binary and ternary synthetic dye combinations on a variety of textiles [2–4], but similar studies on natural dyes are scarce. Some such research has although been reported on cotton [1], jute [5] and wool [6].

Compatibility between two dyes in a binary mixture is a very important aspect for obtaining compound shades on textiles, leather etc. and the use of incompatible dyes in a mixture for compound shades is not advisable. Additionally, since the colour build-up of one dye and the colour build-up of the second dye in a binary mixture occur at different rates, it is difficult to control the development of compound shades to a desirable level to obtain a particular tone of colour. Thus, dyes having comparable or identical exhaustion properties are deemed to be compatible with one another [7]. There are several approaches of judging compatibility between two dyes in a binary/tertiary mixture [1, 4, 8–11].

Though some techniques of determining the compatibility of dyes rely on the assumption that dyes are non-interactive and their rate of dyeing do not change in the presence of another dye, this supposition is not always true. The compatibility between two dyes can be assessed by the following methods [1–5]:

a. By determining the rate of dyeing of individual dyes.

b. By colourimetric analysis of progressive development of shade depth (colour build-up) under varying dyeing conditions i.e., either by varying dye concentration in the binary mixture or by varying time and temperature profiles keeping the dye concentration fixed through plots of K/S vs. ΔL and ΔC vs. ΔH.

c. By determining the diffusion coefficient of the dye by each dye in a pair of binary mixture.

All these three methods are subjective, highly time-consuming, analytical and skill-based requiring a precision control of time and temperature for progressive shade build-up. Additionally, they lack quantitative assessment of degree of compatibility rating.

Thus, it is preferable to use a compatibility test method in which the substrate is dyed with a mixture of dyes under realistic dyeing conditions as opposed to dyeing them with individual dyes and then predicting the compatibility from the rate of dyeing or other dyeing results because all dyeing process variables can affect the final dyeing results/outcomes [4]. The compatibility between two dyes can be assessed by a newer relative compatibility rating method based on the difference in colour difference index for dyeing the substrate with varying proportion of two dyes and a relative grading of degree of compatibility have been suggested [1, 5, 12]. Relative compatibility rating (RCR) method for binary dye mixtures is a more recent method that provides an easily usable way of determining the compatibility of binary combinations of dyes [1].

Previously established [1], this faster and more straightforward RCR method of dye compatibility testing is based on data for CDI differences ($CDI_{Max} - CDI_{Min}$) on dyeing results of varying proportion of individual dyes in a binary mixture following standard/optimised dyeing conditions and method. This newer RCR method can be used to test dye compatibility more quickly than the conventional method, which requires time-consuming, laborious and cumbersome plotting of different colour parameters for two separate sets of dyeing and the determination of the more difficult dyeing coefficient or rate of dyeing. The conventional method also does not provide relative quantitative grading for degree of compatibility.

The present study thus explores the application of newer RCR method of compatibility test of any two dyes, for choosing right compatible dyes only for development of newer shades using binary mixtures of natural dyes (babul bark with annatto seeds, babul bark with Himalayan rhubarb roots and babul bark with pomegranate rind) and the compatibility of the dyes used in the binary mixtures in different proportions on bleached pre-mordanted cotton fabric by both the conventional method of progressive colour build-up (with the plotting of K/S vs. ΔL and ΔC vs. ΔL) as well as the newer relative compatibility rating technique. Both methods have been compared for babul bark with annatto seeds, babul bark with Himalayan rhubarb roots and babul bark with pomegranate rind etc. to judge their efficacy for further endorsement.

2. Methodology

2.1 Fabric

Hand-spun and hand-woven, bleached 100% khadi cotton fabric from Khadi Silk Emporium, Kolkata was used in this study (**Table 1**).

2.2 Natural dyes, chemicals and auxiliaries

Dried gallnuts (*Quercus infectoria*) as a natural mordant, and natural dyes like dried babul bark (*Acacia nilotica*), annatto (*Bixa orellana*), dried pomegranate rind (*Punica granatum*) were obtained from Kangali Charan & Sons, Kolkata. Ready to use natural dye powder of *Rheum emodi* (Himalayan rhubarb) was sourced from AMA Herbal Laboratories Pvt. Ltd., Lucknow.

Gallnuts (*Quercus infectoria*) are collected from the oak tree or similar shrubs like the walnut tree after they are formed as a response to insect infestation. The galls from *Quercus infectoria* contain the highest level of naturally occurring tannins, approximately 50–70% [13]. They also contain 2–4% each of gallic and ellagic acid. The main colouring component in gallnut extract is ellagic acid [14], which has an affinity for dyeing substrates due to the presence of OH (auxochrome group) (**Figure 1**).

The pod/leaves and bark of the Acacia Nilotica (babul bark) tree locally known as *kikar* or babul or Indian gum Arabic tree in India are rich in polyphenols being mainly composed of natural tannins containing gallic acid, ellagic acid and gallotannins having

Fibre	Weave	Warp/ends	Weft/picks	GSM	Thickness
Cotton	Plain	86 ends of 15 tex/inch	61 picks of 16 tex/inch	76 g/m^2	0.31 ± 0.2 mm

Table 1.
Fabric specifications.

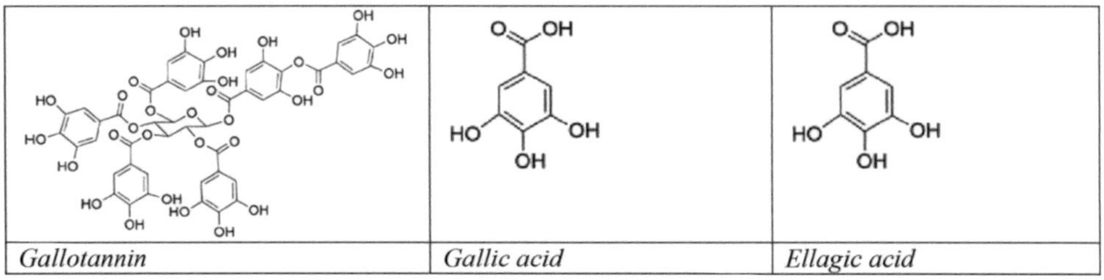

Figure 1.
Chemical composition of gallnut (Quercus infectoria).

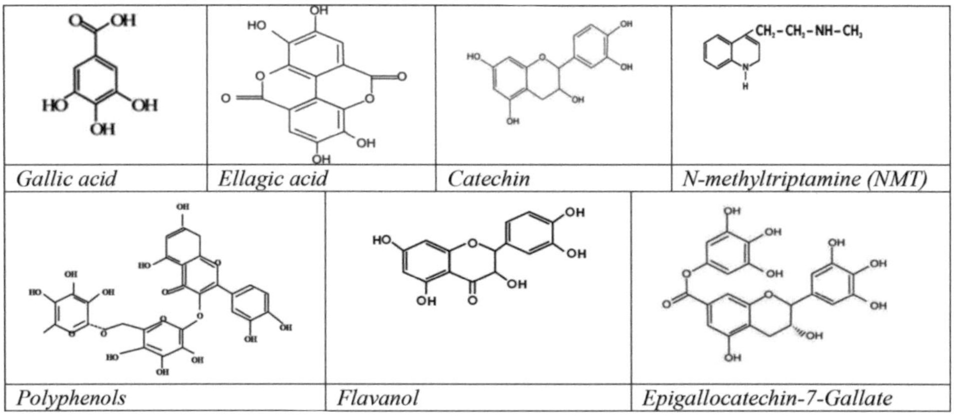

Figure 2.
Chemical composition of babul bark (Acacia nilotica).

light yellow colour in condensed non-hydrolysable forms, some catechin and epigallocatechin gallate having much darker yellow-brown colour gallate [15–17] besides few yellow-coloured amino acids (like NMT) and other minor constituents. The tannin content in babul bark ranges from 9 to 16.5% on dry weight (**Figure 2**) [17, 18].

Bixa orellana (annatto) seeds is a tall, perennial shrub which is the source of 70% of all natural colourants consumed worldwide. The dye contains 80% of coloured unsaturated pigment i.e. bixin (monomethyl ester of the dicarboxylic carotenoid compound), and norbixin (9′-cis-6,6′-diapocarotene-6,6′dioic acid, $C_{24}H_{28}O_4$) as a yellow pigment [19]. In addition to bixin and norbixin, annatto seeds also contain isobixin, bixein, lutein, beta-carotene, cryptoxanthin, crocetin, zeaxanthin, bixol, ishwarane, threonine, tryptophan, phenylalanine, ellagic acid, salicylic acid and tomentosic acid (**Figure 3**) [20].

Rheum emodi (Himalayan rhubarb) is a powerful perennial herb having strong roots and a hollow stem with reddish streaks and a green colour. A variety of anthraquinone derivatives, including emodin, emodin-3-monomethyl ether, chrysophanol, aloe-emodin, and rhein, are present in Indian rhubarb [21] and physcion. Chrysophanolalso known aschrysophanic acid is the main colouring agent of rhubarb (**Figure 4**) [22].

Punica granatum (Pomegranate rind), is the most significant of two species of the Punica Genus that contains a significant amount of tannin in its flower, leaf, root-bark, fruit and peels [23]. The main colouring agent in pomegranate rind is granatonine, which is found as N-methyl granatonine [24]. The fruit's rinds also contain rutin and quercetin. Condensed tannins make up 26% of the chemical content of the peel (**Figure 5**).

DOI: http://dx.doi.org/10.5772/intechopen.114123

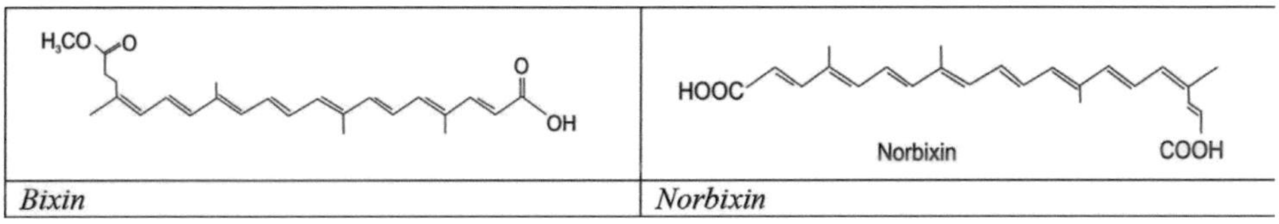

Figure 3.
Chemical composition of annatto seeds (Bixa orellana).

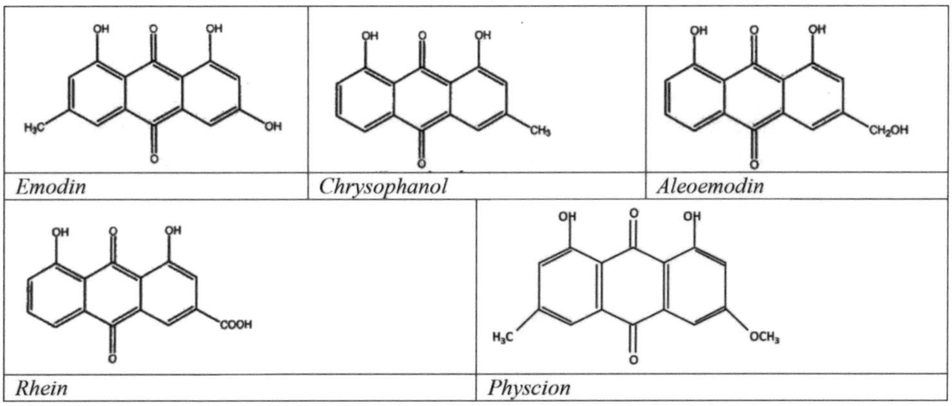

Figure 4.
Chemical composition of Himalayan rhubarb roots (Rheum emodi).

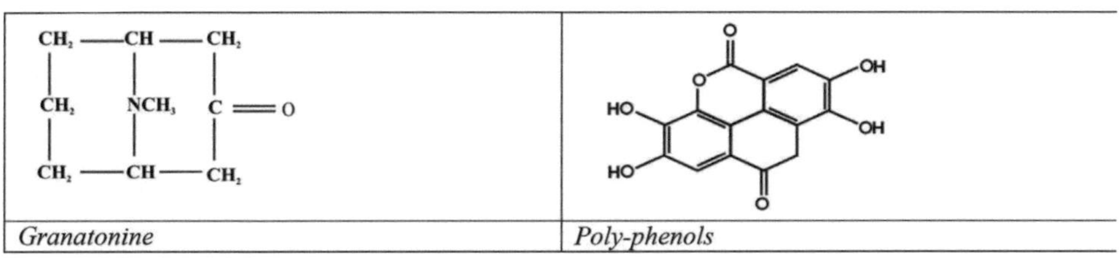

Figure 5.
Chemical composition of pomegranate rind (Punica granatum).

Other chemicals used like amylase enzyme (for desizing), acetic acid and sodium carbonate (for adjusting the pH), conc. sulphuric acid, hydrochloric acid, ferric chloride, aluminium chloride, chloroform, ammonia, and 1% picric acid were obtained from E-Merck (India). Reagents (Dragendroff, Mayer and Wagner) were obtained from Nice Chemicals Pvt. Ltd., Kolkata and non-ionic wetting agent (Axel NW 100) from Bharati Chemicals (Kolkata).

3. Methods

3.1 Pre-treatment of cotton fabric

The cotton fabric was desized using 2 ml/l of amylase enzyme, 5 g/l NaCl, 2 g/l non-ionic wetting agent, 1:30 MLR, at 50–90°C for 15 min as reported in earlier literature [25]. The treated samples underwent a thorough 15-min wash in hot water. Finally, 100°C was used to dry the textile samples to a consistent weight.

3.2 Extraction of colourants from natural resources

All materials from natural sources (gallnut, babul bark, annatto seeds and pomegranate rind) were dried in the sun and powdered using a mechanical grinder.

The aqueous extraction of gallnut powder was done at the optimised conditions reported earlier [26] at MLR—1:20, extraction time—45 min, temperature—80°C and at pH—11.

The aqueous extraction of babul bark was done at the optimised conditions reported earlier [16] at MLR—1:30, extraction time—45 min, temperature—60°C and pH—6.

Aqueous extraction of the colourants from powdered bixa and pomegranate rind were carried out under varying process conditions—pH (3–11), MLR (1:10–1:50), time (15–120 mins) and temperature (37–100°C) and optimization parameters identified based on the optical density with the greatest value at maximum absorbance wavelength of the solution (**Table 2**).

Ready-to-use dye powder of Himalayan rhubarb roots was extracted at 70–80°C for 60 min as recommended by the manufacturer.

3.3 Mordanting of cotton fabric with gallnut extract

Desized cotton fabric was pre-mordanted with gallnut under optimised conditions of mordant concentration—20% owf, 1:20 MLR, 11 pH at 80°C for 45 min as reported earlier [26].

3.4 Dyeing of pre-mordanted cotton using a mixture of two natural dye extracts

Gallnut extract (aqueous) pre-mordanted cotton fabric was dyed with aqueous extracts of both single or selected binary pairs of natural dyes in varying proportions (100:0, 75:25, 50:50, 25:75 and 0:100) as mentioned below:

a. B1: babul bark + annatto seeds (BB & AS)

b. B2: babul bark + Himalayan rhubarb roots (BB & HR)

c. B3: babul bark + pomegranate rind (BB & PR)

The binary mixtures of the dye extracts were applied on gallnut pre-mordanted cotton fabric maintaining the overall total concentration of both the selected dyes (in the binary mixture) at 40% (on the weight of source material) at 100°C for 60 min using MLR 1:20 and 10% owf of sodium chloride as additive using the exhaust process. Following dyeing, the samples were soaked in a 2 g/l soap solution for 15 min at 60°C, then rinsed and allowed to air dry in the shade.

Natural materials	λ_{max}	pH	MLR	Time (in min)	Temperature (°C)
Annatto seeds	410	11	1:20	75	80
Pomegranate rind	420	11	1:20	45	80

Table 2.
Optimised extraction condition parameters for different natural materials.

4. Test methods

4.1 Phytochemical analysis

Analysis of the phytochemicals found in the aqueous and ethanolic extracts of natural mordants and dyes was done using the standard Trease and Evans and Harbone [27, 28] procedure.

4.2 UV-vis spectral analysis

The wavelength of maximum absorbance for 0.1% aqueous extracts of the natural dyes were identified at 410 nm for gallnut and bixa seeds; 420 nm for pomegranate rind; 430 nm for *Rheum emodi* and 510 nm of babul bark, separately, through UV-vis spectral analysis using UV-vis absorbance spectrophotometer (Hitachi-U-2000 model).

4.3 Surface colour strength

Surface colour strength of the control and dyed cotton samples was estimated using the Kubelka Munk equation [29] utilising a Premier Colour Scan reflectance spectrophotometer (model SC 5100A) and related colourlab plus colour matching software to measure the surface reflectance of each of the dyed samples at its respective λ_{max}.

From the measured K/S values for 100% individual dye applied on a pre-mordanted cotton fabric, the calculated K/S values (as K/S values are additive and liner) were obtained for samples dyed with specific properties of a selected pair of dyes using the following equation [5]:

$$\text{Calculated K/S}^{a} = \frac{m}{100}\text{K/S}^{b} + \frac{N}{100}\text{K/S}^{c} \quad (1)$$

Where (^{a}) is observed for the m:n proportion of the two dyes (A & B) in the binary mixture, (^{b}) is observed for 100% of dye A and (^{c}) is observed for 100% of dye B.

4.4 Colour interaction parameters

Using a Premier Colour Scan (model SC 5100A) reflectance spectrophotometer and associated Colourlab plus colour matching software, values for total colour difference (ΔE), lightness/darkness (ΔL^*), redness/greenness (Δa^*), blueness/yellowness (Δb^*), change in chroma (ΔC^*), and change in hue (ΔH_{ab}) were measured before and after dyeing to compare the shade depth and colour differences of each dyed sample against a particular undyed (bleached/mordanted) standard sample [29].

4.5 General metamerism index

Nimeroff and Yurow's equation [30] was used to calculate the general metamerism index (MI) of the treated and dyed samples.

4.6 Colour difference index

The colour difference index (CDI) [1] was also used in the current work to understand the combined effects on total differences in colour strength with varying rate of colour build up and difference in hue and chroma and metamerism for different proportion of dyes or variations in different dyeing process variables by a single index value. CDI indicates the combined effects of various known individual colour difference parameters between any two samples when dyed with varying shade under various dyeing conditions. Relatively, higher CDI values indicate more criticality in the control of colour variation with a particular dyeing condition/variation in the individual dye proportion in their binary mixture for development of compound shades.

Thus, colour difference index (CDI) values represent the overall effect of colour variation arising due to variations in dyeing process variables in terms of the major colour difference parameters between sample-1 (standard) and sample-2 (produced) when dyed under different conditions using different proportion of binary mixture of dyes etc. This indicates overall dispersion of varied colour yield and colour differences in terms of hue, chroma, metamerism values. Colour difference index (CDI) was calculated by an established empirical relationship established earler [1]:

$$\text{Colour Difference Index}(CDI) = \frac{\Delta E \times \Delta H}{\Delta C \times MI} \tag{2}$$

4.7 Compatibility tests for selected binary pairs of natural dyes

4.7.1 Conventional method

The following selected binary pairs (50:50) of natural dyes were applied on the pre-mordanted cotton fabric using an overall 40% (owf) of the respective extracts.

Two sets of gallnut pre-mordanted cotton fabric samples were dyed in progressive depth of shade for each selected binary pair of dyes (babul bark with annatto seeds, Himayalan rhubarb roots, and pomegranate) taken in equal proportions (50:50).

B1: babul bark + bixa seeds (BB & AS).

B2: babul bark + Himalayan rhubarb roots (BB & HR).

B3: babul bark + pomegranate rind (BB & PR).

The shade was gradually/progressively developed in the first set of samples (Set-I) by adjusting the dyeing time and temperature. Three separate small pre-mordanted cotton fabric samples were dyed for each pair of dyes, using Lab Dyer MAG Solvics make with a temperature controller for different dyeing time periods (15–60 min). At intervals of 15 min starting at 60°C, the samples were taken out of the dye bath while still being heated at a rate of 1° per min until 90°. The first sample was removed after 15 min at 60°C, and the last one was removed 60 min later at 90°C, marking the end of the dyeing process. The incremental depth of darkness (progressive shade depth) in Set-II was created by changing the dye mixture's overall concentration from 10 to 40%. Three separate small pre-mordanted cotton fabric samples were dyed in triplicate for each pair of dyes at intervals of 10% using 10–40% (owf) of each pair of dyes applied in equal proportion (50:50) at 90°C for 60 min.

The dyed fabrics for both dyeing processes (Set-I and Set-II) were washed, soaped, and rinsed before being air-dried. The differences in the CIELab coordinates, ΔL,

Δa, Δb and ΔC for all the dyed fabrics using Set I and II methods in comparison to the standard undyed fabric sample were measured separately using the Macbeth 2020 Plus Reflectance Spectrophotometer and related colour matching software. The compatibility of a selective pair of dyes was judged from the degree of closeness and overlapping of the two curves i.e., ΔC vs. ΔL or K/S vs. ΔL corresponding to the two sets of dyeing (Set-I & Set-II).

4.7.2 Newer RCR method (alternative)

The related ΔE, ΔC, ΔH and MI values' magnitudes, regardless of their sign or direction, may be utilised to produce the CDI (colour difference index) after the application of various ratios of binary pairs of dyes on the same fabric [4]. The compatibility rating (between 0 and 5, rating 5 shows maximum or excellent compatibility, rating 1 represents minimal or worst compatibility, and rating 0 is viewed as fully non-compatible) increases with the proximity of the CDI values for binary pairings of dyes.

4.8 Colour fastness properties

Fastness to washing of the dyed cotton was assessed using the Launder O Meter in accordance with AATCC Test Method 8-2009 [31]; fastness to light was assessed using AATCC Test Method 16-2004 [31]; fastness to rubbing (dry and wet) was evaluated using AATCC Test Method 8-2007 [31]; and fastness to perspiration (alkaline and acidic) was assessed using AATCC Test Method 15-2009 [31].

5. Results and discussions

5.1 Standardisation of extraction conditions for annatto seeds and pomegranate rind

The colouring matter from annatto seeds and pomegranate rind was extracted under various pH, MLR, time, and temperature conditions. The optical values for each parameter were established based on the highest optical densities of the corresponding solutions (**Table 3**).

Aqueous extraction of annatto seeds under alkaline pH gave a higher colour yield as shown in **Table 2**. pH 11 gave the highest optical density at 410 nm. A higher liquor ratio probably has a dilution effect thereby lowering the optical density of the extract. Thus, an MLR of 1:20 was considered optimum. The optical density of annatto seeds shows an increasing trend till 75 min after which the remains the same. Hence 75 min is considered optimum. The highest optical density was observed at 80°C, after which it declined. This indicates that most of the colouring matter from the annatto seeds is leached out at this temperature. Thus 80°C temperature is considered optimum.

Extraction of pomegranate rind is most favourable in an alkaline medium and gives better colour yield with the highest optical density value obtained at pH 11 (**Table 3**). Low MLR of 1:20 and 1:30 are best suited for extraction of pomegranate rind, but 1:30 requires a high amount of water and energy; hence, 1:20 was selected as optimum. After 45 min the optical density reduces, thus 45 min is considered optimum. With the increase in temperature to 80°C, the optical density diminishes. Thus, a temperature of 80°C was considered optimum (**Table 4**).

Extraction process condition		Optical density at λ_{max}		
		Gallnut (at 410 nm)	Annatto seeds (at 410 nm)	Pomegranate rind (at 420 nm)
pH	3	1.2	0.5	1.6
	5	1.2	1.0	1.6
	7	1.3	1.6	1.7
	9	1.3	1.7	1.7
	11	1.4	1.8	1.8
MLR	1:10	1.1	1.8	1.7
	1:20	1.2	1.9	1.8
	1:30	1.1	1.6	1.8
	1:40	1.0	1.2	1.7
	1:50	0.9	1.1	1.6
Time	15	1.2	1.0	1.4
	30	1.2	1.3	1.6
	45	1.3	1.5	1.7
	60	1.2	1.6	1.6
	75	1.2	1.8	1.6
	90	1.2	1.8	1.5
	120	1.2	1.8	1.5
Temp	RT	1.1	1.1	1.5
	60	1.2	1.6	1.6
	80	1.3	1.8	1.7
	100	1.1	1.7	1.6

Table 3.
Optical densities at λ_{max} of aqueous extract of gallnut, annatto seeds and pomegranate rind.

Natural resources	pH	MLR	Time (min)	Temperature (°C)
Gallnut	11	1:20	45	80
Annatto seeds	11	1:20	75	80
Pomegranate rind	11	1:20	45	80
Himalayan rhubarb roots	—	—	60	70–80

The extraction of Himalayan rhubarb roots was done as recommended by the manufacturer.

Table 4.
Optimised extraction conditions for aqueous extract of gallnut, annatto seeds and Himalayan rhubarb roots.

5.2 Phytochemical analysis of different natural resources

Gallnut, babul bark, annatto seeds, Himalayan rhubarb roots and pomegranate rind were extracted in two media—aqueous and ethanolic and phytochemical screening was done for each extract (**Table 5**).

Phytochemical screening	Gallnut		Babul bark		Annatto seeds		Himalayan rhubarb roots		Pomegranate rind	
	Aq	Et	Aq	Et	Aq	Et	Aq	Et	Aq	Et
Alkaloids										
Dragendorff's reagent (*potassium bismuth iodide*)	2+	2+	2+	3+	3+	3+	2+	2+	2+	3+
Mayer's reagent (potassiomercuric iodide)	1+	2+	2+	2+	2+	2+	1+	2+	2+	2+
Wagner's reagent (*iodo-potassium iodide*)	2+	3+	2+	3+	3+	3+	2+	2+	2+	3+
Picric acid [$(O_2N)_3C_6H_2OH$]	2+	3+	2+	3+	3+	3+	2+	2+	3+	3+
Flavonoids										
Ammonium	2+	3+	2+	3+	3+	3+	3+	3+	3+	3+
Aluminium chloride	2+	3+	2+	3+	3+	3+	3+	3+	3+	3+
Tannins										
Ferric chloride	2+	3+	2+	3+	1+	1+	1+	1+	2+	3+
Terpenoids										
Salkowski	2+	2+	1+	3+	3+	3+	3+	3+	2+	3+
Saponins										
Emulsion	2+	2+	2+	2+	—	—	—	—	1+	1+

Aq, aqueous extract; Bt, ethanolic extract; –, absence of compound; +, presence in low amount; ++, presence in moderate abundance; +++, presence in high amount.

Table 5.
Qualitative examination of the phytochemicals in gallnut, babul bark, annatto seeds, Himalayan rhubarb roots and pomegranate rind.

The phytochemical screening of gallnut, babul bark, annatto seeds, Himalayan rhubarb roots and pomegranate rind extracts indicated the existence of certain compounds in their extracts (extraction in ethanol was done only for phytochemical analysis). Flavonoids, alkaloids, tannins, terpenoids and saponins are present in all five extracts prepared under different mediums. The extract of gallnut, babul bark and Himalayan rhubarb roots showed a moderate presence of alkaloids in both the mediums. Flavonoids were present in high amounts in the extraction extract of all five compounds, but it was present in moderate amounts in the aqueous extract. Tannins were present in high amounts when extraction was carried out in an ethanolic medium for gallnut, babul bark and Himalayan rhubarb roots, but it was present in moderate amounts in aqueous extract. Terpenoids were also present in high amounts in both the mediums (aqueous and ethanol) in bixa seeds and Himalayan rhubarb

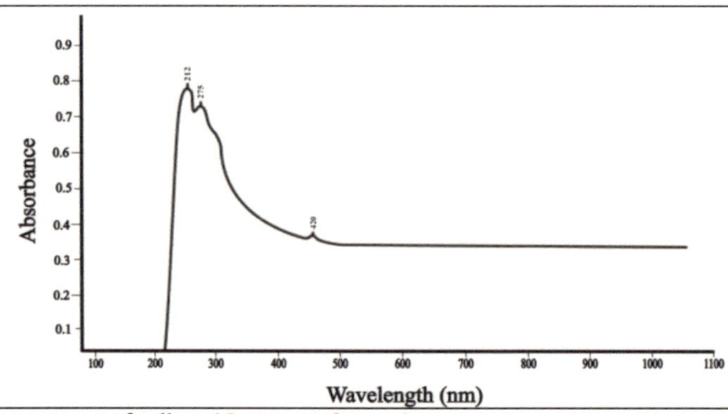

(a) Aqueous extract of gallnut (*Quercus infectoria*)

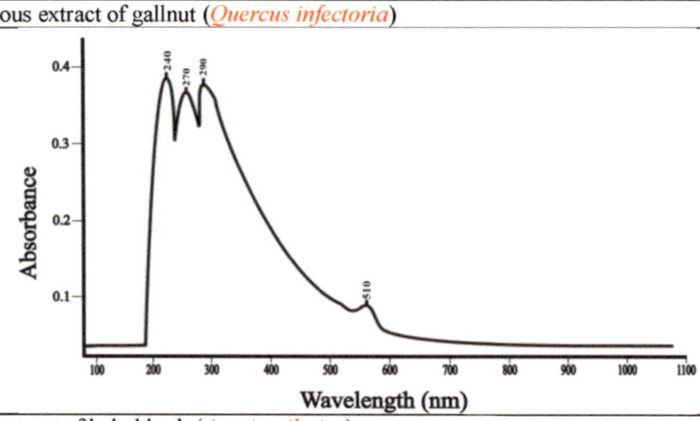

(b) Aqueous extract of babul bark (*Acacia nilotica*)

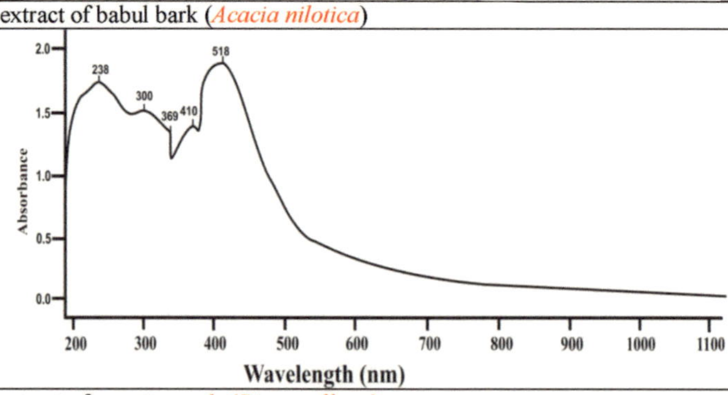

(c) Aqueous extract of annatto seeds (*Bixa orellana*)

DOI: http://dx.doi.org/10.5772/intechopen.114123

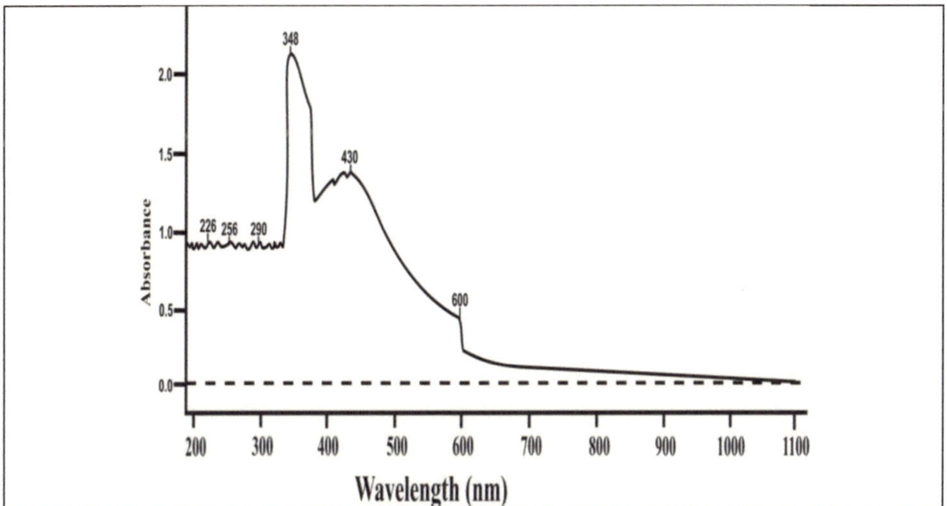

(d) Aqueous extract of Himalayan rhubarb roots (*Rheum emodi*)

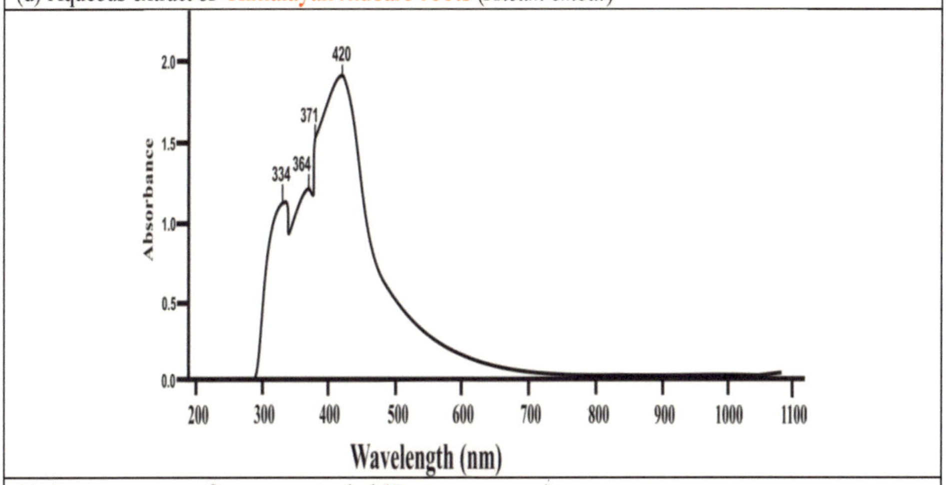

(e) Aqueous extract of pomegranate rind (*Punica granatum*)

Figure 6.
(a–e) Absorbance spectra of gallnut, babul bark, annatto seeds, Himalayan rhubarb roots and pomegranate rind extracts in the UV-range (100–1100 nm).

roots whereas in gallnut, babul bark and pomegranate rind is present in moderate amounts. Saponins were present in considerable amounts in gallnut, babul bark and pomegranate rind in both the extracts, but bixa seeds and Himalayan rhubarb roots do not report the presence of it.

5.3 UV-vis spectral analysis and UV-vis characterisation

UV-vis spectra with the corresponding UV-vis peak frequency at different wavelength (nm) of gallnut, babul bark, annatto seeds, Himalayan rhubarb roots and pomegranate rind aqueous extracts are shown in **Figure 6(a–e)** and characterisation of the different peaks in the visible and UV zone have been tabulated (**Table 6**).

Sl. no.	Name of the natural source (aqueous extract)	Peaks in UV zone (nm) & content		Peaks in visible zone & content	
1	Gallnut	212	Gallic acid and gallotannins	420	Gallic acid and gallotannins
		275	Gallic acid and gallotannins	—	
2	Babul bark	240	Condensed gallotannins	510	Dark brown coloured catechin and epigallocatechin gallate
		270	Condensed gallotannins	—	—
		290	Condensed gallotannins	—	—
3	Annatto seeds	238	Careteniod isomers	410–420	Bixin
		370	Some flavonoid and phenolic compounds	490–520	Nor-bixin
4	Himalayan rhubarb	348	Phenolic componds	430–450	Anthraquinone derivatives
		290	Condensed gallotannins	550–600	Anthraquinone compounds
		256–226	Hydroxy anthroquinone	—	—
5	Pomegranate rind	371	Presence of flavonoid and phenolic compounds	420	Presence of gallotannins
		364	Presence of C=O group	—	—
		334	Presence of phenolic compounds	—	—

Table 6.
Identification and interpretation of peaks in the visible and UV zones.

5.3.1 Bixa orellana (annatto) seeds

Three peaks have been identified at 240 nm, 416 nm and 494 nm confirming the presence of carotenoid compounds [32]. Other researchers have [33] also found out, that the annatto seeds extract exhibits the highest absorbance between 400 and 500 nm (λ_{max}), confirming the presence of bixin and nor-bixin compounds.

5.3.2 Rheum emodi (Himalayan rhubarb) Roots

Absorption peaks in aqueous extract of Himalayan rhubarb roots (Rheum emodi) were identified at 226 nm, 256 nm, 290 nm, 348 nm, 430 nm and 600 nm. The absorption peaks around 226–256 nm indicate the existence of hydroxyanthraquinone group [34], absorption at 290–340 nm may be due to the occurrence of several phenolic compounds [35] while in the visible range, the absorption peaks of 430 nm and 600 nm show the presence of anthraquinone compounds and derivatives [36].

5.3.3 Punica granatum (Pomegranate) rind

It shows a significant peak between 330 and 370 nm confirming the presence of flavonoid and phenolic compounds [37]. A significant peak in the visible region 420 nm is visible and it indicates the presence of gallotannins [38].

5.4 Colour strength, related parameters and brightness index of pre-mordanted cotton fabric with gallnut dyed with selected binary pairs of natural dyes in different proportions

Colour strength, related parameters and brightness index of pre-mordanted cotton fabric with gallnut dyed with selected binary pairs of natural dyes in different proportions namely B1—babul bark + annatto seeds (BB & AS), B2—babul bark + Himalayan rhubarb (BB & HR) and B3—babul bark + pomegranate rind (BB & PR) are tabulated in **Table 7**.

The observed K/S (surface colour strength) of cotton dyed with a mixture of babul bark and annatto seeds (BB + AS), Himalayan rhubarb (BB + HR), and pomegranate rind (BB + PR) is slightly lower than the theoretical K/S that has been calculated separately by adding the K/S for each proportion of the dye in the binary mixture (**Table 7**). This is true for all different ratios of the two dyes used in the binary mixture (i.e., 25:75, 50:50, and 75:25) although the difference varies from one pair of dyes to another and for different ratios utilised.

The metameric effect considering the differences in the two K/S values measured at λ_{max} of the respective dyes in the binary pair (D2) for each pair of natural dyes is found to be minimum for the B2 (BB + HR) mixture. The order of increasing

Dye combination	Proportion of binary pairs of dyes									
	100:0		75:25		50:50		25:75		0:100	
	A_{D1}	B_{D2}	A_{D1}	B_{D2}	A_{D1}	B_{D2}	A_{D1}	B_{D2}	A_{D1}	B_{D2}
B1 (BB:AS)										
Observed	5.5	6.6	5.3	6.9	5.4	6.8	5.5	6.7	6.1	6.9
Calculated	—	—	5.6	6.7	5.7	6.8	5.9	6.8	—	—
D1	—	—	0.3	0.2	0.3	0.0	0.4	0.1	—	—
D2	1.1		1.6		1.4		1.2		0.8	
B2 (BB:HR)										
Observed	5.5	6.1	5.1	5.4	4.7	4.8	4.4	4.5	4.5	4.7
Calculated	—	—	5.2	5.7	4.9	5.3	4.8	4.9	—	—
D1	—	—	0.1	0.3	0.2	0.5	0.4	0.5	—	—
D2	0.6		0.3		0.1		0.1		0.2	
B3 (BB:PR)										
Observed	5.5	6.3	5.6	5.3	4.7	5.1	4.5	5.0	4.4	4.3
Calculated	—	—	5.3	5.8	4.9	5.2	4.7	4.8	—	—
D1	—	—	0.3	0.5	0.2	0.1	0.2	0.2	—	—
D2	0.8		0.3		0.4		0.5		0.1	

BB, babul bark; AS – Annatto seeds; HR – Himalayan rhubarb roots; PR, pomegranate rind; AD1, K/S at λ_{max} of BB; BD2, K/S at λ_{max} of BX/RE/PR; D1, difference between observed and calculated values; D2, difference between two sets of observed vales at the respective λ_{max} of the two dyes in the binary pair.

Table 7.
Observed and calculated K/S values and related parameters of cotton fabric pre-mordanted with gallnut and dyed with selected binary pairs of natural dyes in different proportions.

difference between two sets of observed K/S values at two different wavelengths values for each pair of natural dyes is given in **Table 7**:

$$B2(BB+RE) < B3(BB+PR) < B1(BB+BX)$$

The surface colour strength for all binary pairs of dyes is highest for B1 (BB + AS) compared to other binary pairs B2 (BB + HR) and B3 (BB + PR), when K/S is assessed at the common wavelength of all the dyes used (410 nm). K/S values' ascending order for the different dyes used in varying ratios in their binary pairs is

$$B1(BB+BX) < B3(BB+PR) < B2(BB+RE)$$

K/S reduces with decreasing amount of babul bark in the mixture for B2 (BB + HR)) and B3 (BB + PR) binary pairs (**Table 8**), but for the B1 (BB + AS) pair of dyes, the presence of more babul bark in the mixture decreases the surface colour strength (K/S) in general.

Dyeing conditions	Proportion	K/S at λ_{max} (410 nm*)	ΔE	ΔC	ΔL	ΔH	BI	MI
Control (gallnut pre-mordanted cotton)		3.9	—	—	—	—	—	—
B1 (BB + AS)	100:0	6.6	18.6	9.9	15.2	9.9	10.4	3.8
	75:25	6.9	19.3	9.1	13.0	10.9	10.2	5.0
	50:50	6.8	19.5	12.2	10.8	10.8	10.4	5.3
	25:75	6.7	23.1	17.6	9.8	11.3	9.4	6.2
	0:100	6.9	37.9	23.8	10.4	13.6	5.2	7.7
B2 (BB + HR)	100:0	6.5	19.1	8.4	15.8	9.9	10.3	5.0
	75:25	5.8	17.2	4.8	13.9	8.9	10.6	3.8
	50:50	4.3	11.9	4.0	9.1	6.7	13.1	3.5
	25:75	3.6	6.9	3.8	7.5	3.7	16.0	2.8
	0:100	3.5	5.9	3.7	6.8	1.5	17.7	1.9
B3 (BB + PR)	100:0	6.6	18.2	7.9	14.8	9.8	10.5	4.9
	75:25	6.7	12.9	6.0	9.5	6.5	13.5	3.6
	50:50	5.9	7.3	5.4	7.3	3.6	17.9	2.6
	25:75	4.7	8.8	7.9	6.3	2.4	16.2	1.5
	0:100	4.4	6.6	6.0	5.3	1.1	18.1	1.1

*BB, babul bark; AS – Annatto seeds; HR – Himalayan rhubarb roots; PR, pomegranate rind; Control, cotton fabric pre-mordanted with gallnut extract; ΔC, change in chroma; ΔH, change in hue; ΔE, total colour difference; MI, metamerism index; BI, brightness index. *410 nm—common.*

Table 8.
Colour strength, metamerism and brightness index and other colour related parameters of cotton fabric pre-mordanted with gallnut and dyed with selected binary pairs of natural dyes in different proportions.

Colour difference (ΔE) increases with a reduction in the proportion of babul bark in the binary mixture with annatto seeds (B1), but it reduces with an increase in babul bark proportion in the mixture with Himalayan rhubarb (B2) or pomegranate rind (B3). The binary pair of babul bark and pomegranate rind (B3) shows the lowest ΔE regardless of the ratio of each dye used in the binary mixture (**Table 8**). This is followed by B2 (BB + HR) and B1 (BB + AS) pairs of dyes. The increasing order of ΔE values for the three selected binary pairs of dyes are in the following order:

$$\mathrm{B1}(BB+BX)<\mathrm{B2}(BB+RE)<\mathrm{B3}(BB+PR)$$

Change in chroma (ΔC) is positive indicating that it is lowest or minimum for all corresponding dye proportions (25:75, 50:50 and 75:25) used in the B2 (BB + RE) pair of dyes, while a maximum change in ΔC is observed for B1 (BB + AS) pair of dyes used in different ratios (**Table 8**). The increasing ΔC values for the three selected binary pairs of dyes are in the following order:

$$\mathrm{B1}(BB+BX)<\mathrm{B3}(BB+PR)<\mathrm{B2}(BB+RE)$$

The B1 (BB + AS) pair of dyes has the highest growing order of magnitude for ΔH values across all ratios/combinations (**Table 8**). The increasing ΔH values for the three selected binary pairs of dyes are found to be in the following order:

$$\mathrm{B1}(BB+BX)<\mathrm{B2}(BB+RE)<\mathrm{B3}(BB+PR)$$

Brightness index (BI), another important colour characteristic for dyed materials, is significantly impacted by surface sheen and specular reflectance. The brightness index values for the selected binary pair of dyes used in this study are found to be in the following order (**Table 8**):

$$\mathrm{B3}(BB+PR)<\mathrm{B2}(BB+RE)<\mathrm{B1}(BB+BX)$$

Metamerism index (MI) is minimum for B3 (BB + PR) pair of dyes under all different ratios used in the binary pair. The order of increasing MI for the selected binary pair of dyes used in this study is found to be as follows:

$$\mathrm{B3}(BB+PR)<\mathrm{B2}(BB+RE)<\mathrm{B1}(BB+BX)$$

5.5 Colour fastness of cotton fabric pre-mordanted with gallnut and dyed withselected binary pairs of natural dyes in different proportions

The fastness of cotton fabric pre-mordanted with gallnut and dyed with selected binary pairs of natural dyes in different proportions (**Table 9**) indicates that the light fastness of all combinations of dyes in the selected binary mixture are found to be good (rating of 4 and 5). Wash fastness for all the dyed samples also ranges between moderate (rating of 2–3) to good (rating of 3–4). All of the samples dyed with binary

Dyeing conditions	LF	Wash fastness (ISO-II)			Crocking fastness		Perspiration fastness					
							Acidic			Alkaline		
		LoD	StC	StCW	Dry (STC)	Wet (STC)	LoD	StC	StCw	LoD	StC	StCw
Control*	4	4	5	5	5	4–5	3–4	4	4	3–4	3–4	4
75:25 proportion												
B1 (BB + AS)	5	3–4	4	3–4	4–5	3–4	3	4	4	3–4	4	4–5
B2 (BB + HR)	5	2–3	3–4	3–4	4–5	4	3–4	4	3–4	3	4	4
B3 (BB + PR)	5	3	3–4	3–4	4–5	4	3–4	3–4	3–4	2–3	3–4	3–4
50:50 proportion												
B1 (BB + AS)	5	3	4	3–4	4–5	3–4	3	4	3–4	3–4	4	3–4
B2 (BB + HR)	5	2–3	3–4	4	4–5	4	3–4	4	4	3	3	3–4
B3 (BB + PR)	5	3	3–4	3–4	4–5	4	4	4	3–4	3	3–4	3–4
25:75 proportion												
B1 (BB + AS)	4	3	3–4	3–4	4–5	3–4	3	4	3	3–4	4	4
B2 (BB + HR)	5	3–4	4	3–4	4–5	3–4	3–4	4	3–4	3	4	3–4
B3 (BB + PR)	4	3–4	4	4	4–5	4	3–4	3–4	3–4	3	4	4

BB, babul bark; AS – Annatto seeds; HR – Himalayan rhubarb roots; PR, pomegranate rind; LoD, loss in depth of shade; StC, staining of adjacent cotton fabric; StCw, staining of adjacent cotswool fabric; LF, light fastness. *Cotton fabric pre-mordanted with gallnut extract.*

Table 9.
Colour fastness of cotton fabric pre-mordanted with gallnut and dyed with selected binary pairs of natural dyes in different proportions.

mixtures of natural dyes in various proportions have excellent (rating of 4–5) and good (rating of 3–4) dry and wet rubbing fastness, respectively, indicating that no superficial unfixed dye molecules are left on the fibre surface and the dyes have been well absorbed inside the fibre. For all binary pairs of natural dyes, fastness to acidic perspiration is either comparable to or significantly superior to alkaline perspiration fastness.

5.6 Compatibility tests for selected binary mixtures of natural dyes

Binary dye pairings exhibit a wide range of responses to different dyeing techniques. A pair of dyes may show compatibility under one set of dyeing conditions, but they may turn out to be incompatible under another set of conditions. When a certain fibre is dyed with one dye, it does not necessarily follow that the dye will behave the same way when mixed with another dye.

Plotting K/S vs. L and C vs. L for two sets of progressive depth of shade produced by dyeing pre-mordanted cotton with the selected binary mixture of natural dyes taken in equal proportion, and then judging the closeness and degree of overlap between two sets of curves (Set-I and Set-II) are the traditional/conventional methods for testing compatibility between dyes. To provide dyers the choice of selecting the right mixture of dyes to match a desired compound hue, it is important to test the relative compatibility of various binary mixtures of natural dyes using some type of quantitative term or rating system (**Table 10**).

In the current investigation, two approaches of testing the compatibility of binary dye pairs have been employed. In the conventional method, the closeness and degree of overlap between the two sets of curvesin the plots ΔC vs. ΔL or K/S vs. ΔL produced using two sets of dyeing procedures (Set-I and Set-II) for the progressive build-up of shade have been compared. Also, the compatibility of various pairings of selected natural dyes has been established by an easy quantitative method based on relative compatibility rating (RCR) between any pair of natural dyes as described in the earlier study [5]. As suggested earlier the closer the CDI values of dyeing with different binary pair ratios of dyes, the greater their compatibility (RCR). Thus, the compatibility between the two approaches (traditional and proposed) for binary pair of dyes used in different proportions has been compared.

Figure 7 shows the K/S vs. ΔL (plots a–c) and ΔC vs. ΔL (plots a′–c′) plots for both sets (Set-I and Set-II) of dyed materials for three separate pairs (B1–B3) of natural dyes.

The two curves (Set-I and Set-II) for B1 (BB + AS) binary pair of dyes plots for K/S vs. ΔL of the dyed cotton samples exhibit a similar pattern but with a significant gap between the curves in the initial stage, which narrows and gets closer until it almost touches each other at the end of the dyeing cycle when dyeing equilibrium is reached. They indicated that for the K/S vs. ΔL plot for the Set-I curve, the colour build-up was initially slow at a lower temperature, while for the Set-II curve, it was moderately fast due to the use of higher temperatures (plot a, **Figure 7**). These observed variations in these two sets of curves could possibly be due to the difference in the molecular weights of the two dyes. However, both Set-I and Set-II curves are quite close to one another towards the end at the dyeing equilibrium indicating that the colour build-up is at par after enough time has passed during the dyeing process. In contrast, the two curves for Sets I and II show a similar pattern and run almost parallel on plots (a’) of ΔC vs. ΔL. Plots of the Set-I curve and Set-II curve show a similar rate of colour build-up with minor deviations, with the maximum increase occurring at 40% dye

Dyeing conditions		Time (in min)	Temperature (°C)	Dye concentration (%)	K/S	ΔC	ΔL
B1 (BB + AS) in 50:50							
SET-I (dye concentration fixed at 40% owf with varying time and temperature)	BX1	15	60	40	6.5	11.8	−6.3
	BX2	30	70	40	6.6	11.6	−6.6
	BX3	45	80	40	6.9	11.8	−7.2
	BX4	60	90	40	6.7	12.9	−7.1
SET-II (time and temperature fixed with varying dye concentration)	BX5	60	80	10	6.5	12.1	−6.2
	BX6	60	80	20	6.6	10.4	−6.8
	BX7	60	80	30	7.0	12.9	−7.4
	BX8	60	80	40	6.9	12.5	−7.6
B2 (BB + HR) in 50:50							
SET-I (dye concentration fixed at 40% owf with varying time and temperature)	BR1	15	60	40	4.3	7.1	−5.5
	BR2	30	70	40	4.1	5.5	−5.0
	BR3	45	80	40	4.1	6.7	−5.0
	BR4	60	90	40	4.4	5.8	−4.4
SET-II (time and temperature fixed with varying dye concentration)	BR5	60	80	10	3.9	6.3	−4.6
	BR6	60	80	20	3.9	6.1	−3.9
	BR7	60	80	30	4.2	6.4	−5.0
	BR8	60	80	40	4.4	6.8	−5.8
B3 (BB + PR) in 50:50							
SET-I (dye concentration fixed at 40% owf with varying time and temperature)	BX1	15	60	40	5.1	5.5	−2.5
	BX2	30	70	40	5.6	6.1	−3.3
	BX3	45	80	40	5.7	6.3	−3.6
	BX4	60	90	40	5.9	6.7	−3.7
SET-II (time and temperature fixed with varying dye concentration)	BX5	60	80	10	4.6	4.8	−2.8
	BX6	60	80	20	5.0	5.6	−2.9
	BX7	60	80	30	5.6	5.9	−3.8
	BX8	60	80	40	6.0	6.3	−4.9

BB – Babul bark; AS – Annatto seeds; HR – Himalayan rhubarb roots; PR – Pomegranate rind.

Table 10.
Colour strength and colour related properties of cotton fabric pre-mordanted with gallnut and dyed with selected binary pairs of natural dyes in 50:50 proportion under two sets of dyeing conditions—Set-I under varying conditions of time and 60 temperature, and Set-II under varying dye concentration.

concentration, time of 60 min, and temperature of 90°C. This is likely due to the different major hues of these two dyes, which results in maximum shade developed by sufficient absorption of both babul bark and annatto seed natural dyes. In the case of binary pair B1 (BB + BX), the K/S vs. ΔL and ΔC vs. ΔL plots show that both curves for Sets I and II overlap with slight deviation, indicating a good degree of compatibility. In the proposed RCR method, this pair of dyes exhibits a Grade 3–4 (moderate) relative compatibility rating (**Table 11**).

Thus, the two curves for Set-I and Set-II in the plots of K/S vs. ΔL for the binary mixture B2 (BB + HR) do not exhibit a similar pattern for the build-up of colour from

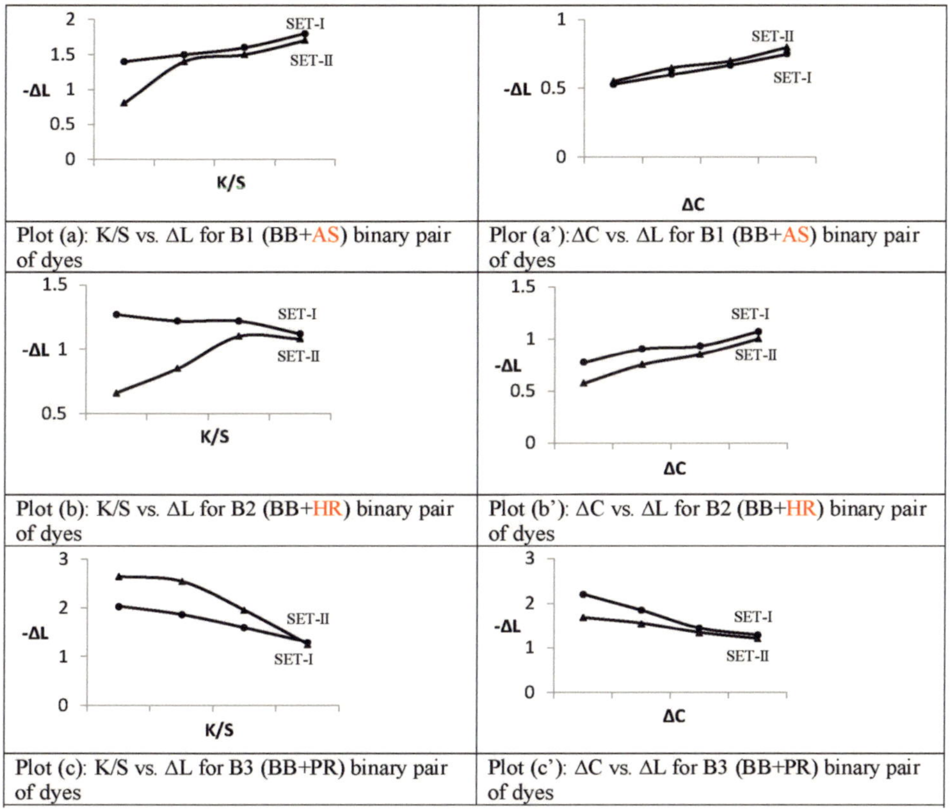

Figure 7.
Plots for K/S vs. ΔL (a–c) and ΔC vs. ΔL (a'–c') for dyeing of three separate pairs (B1–B3) of natural dyes on pre-mordanted cotton fabric.

Dye combination	CDI			$CDI_{max} - CDI_{min}$	RCR	Compatibility grade
	75:25	50:50	25:75			
B1 (BB + AS)	4.2	4.0	3.9	0.3	3–4	Moderate
B2 (BB + HR)	5.5	5.2	4.6	0.9	2–3	Fair
B3 (BB + PR)	5.4	5.1	5.0	0.4	3	Average

BB – Babul bark; AS – Annatto seeds; HR – Himalayan rhubarb roots; PR – Pomegranate rind.

Table 11.
Colour difference index (CDI) and relative compatibility rating (RCR) of cotton pre-mordanted fabric dyed selected binary pairs of natural dyes.

beginning to end. Set-I decreases with an increase in time and temperature and Set-II increases with an increase in dye concentration till 30% owf concentration after which it declines. But for plots (b') of ΔC vs. ΔL, both the two sets show a similar trend.

Because of the closeness of the hues of these two dyes, it may be assumed that the observed gaps between Set-I and Set-II curves in ΔC vs. ΔL plots and K/S vs. ΔL plots become smaller with increasing time, temperature, and dye concentration. Consequently, regardless of the dyeing time and concentrations, the dulling effect does not diminish throughout the entire dyeing cycle in parallel sets, always maintaining a gap. In the case of the newer RCR method, higher CDI values for the binary mixture of natural dyes B2 (BB + HR) are revealed by the presence of chrysophanol and other anthraquinone derivatives, which differ greatly from the gallotannins of babul bark in terms of molecular weight. Therefore, there is little compatibility.

Thus, in the case of binary pair B2 (BB + HR), the plot of K/S vs. ΔL shows the two curves for Sets I and II to be widely spaced initially which gradually overlap with an increase in K/S values. The corresponding plots for ΔC vs. ΔL, also show the same trend, but the gap between the two curves for Set-I and Set-II is much less. Hence, both the set of plots i.e., K/S vs. ΔL and ΔC vs. ΔL for this pair of dyes (B1 – BB + AS) indicate a fair degree of compatibility as also corroborated by the RCR of 2–3 under the more recent relative compatibility classification system (**Table 11**). Thus, both the compatibility results obtained by the traditional method employing ΔC vs. ΔL and K/S vs. ΔL plots and the newer RCR method are similar.

In plots (c) of K/S vs. ΔL and (c') of ΔC vs. ΔL for a binary mixture of B3 (BB + PR), the two curves for Set-I and Set-II exhibit nearly identical trend that increases with an increase in dye concentration, dyeing time, and temperature. The build-up of K/S (depth of shade) and the corresponding darkness (ΔL) is similar (plot c, **Figure 7**) and therefore there is less gap between Set-I and Set-II curves, indicating moderate to fair compatibility between these two dyes in the binary mixture. For the ΔC vs. ΔL plots, the change in chroma (ΔC) and darkness (ΔL) do not build up systematically and change continuously with the increase in dye concentration as well as with the progress in dyeing time almost up to the end, indicating a moderate to average compatibility between this pair of dyes. These two colours are absorbed at various rates and have quite different molecular weights. Additionally, this may be because of the added dulling action of one colourant (gallotannins) in babul bark, which may occasionally be compatible with the colouring matter in pomegranate rind. According to a comparison of the chemical structures of natural dyes, these results may be interpreted as the result of closer molecular weights of the key colour components involved, even though there are significant variances in the dominant hues among them.

Thus, in the case of binary pair B3 (BB + PR), the plot of K/S vs. ΔL shows the two curves for Sets I and II to be widely spaced initially which gradually overlap with an increase in K/S values. The corresponding plots for ΔC vs. ΔL, also show the same trend, but the gap between the two curves representing Set-I and Set-II conditions of dyeing is much less. Hence, both the set of plots i.e., K/S vs. ΔL and ΔC vs. ΔL for this pair of dyes i.e., B1 (BB + AS) indicate a moderate degree of compatibility as also corroborated by the RCR of 3 (**Table 11**).

6. Conclusions

The optimised extraction recipe for annatto seeds is found to be pH 11, MLR 1:20, 75 min time at 80°C temp and for pomegranate rind, it is pH—11, MLR 1:20, 45 min time and 80°C temp. Phytochemical screening indicates the presence of tannins, flavonoids, alkaloids, terpenoids and saponins in the extracts of gallnut, babul bark, annatto seeds, Himalayan rhubarb roots and pomegranate rind. In terms of colour strength and colour related parameters B1 (BB + AS) gave the best results when compared to B2 (BB + HR) and B3 (BB + PR). All samples dyed with binary mixtures of natural dyes in various proportions exhibited good to excellent colour fastness properties.

Thus, comparing to methods of determination of compatibility of three pairs of binary mixtures of natural dyes, it is concluded that the B1 pair (babul bark with bixa seeds) is moderately compatible as per K/S vs. ΔL and ΔC vs. ΔL plots of colour build-up as observed in conventional method; while the newer relative compatibility rating (RCR) method also shows this B1 pair as moderately compatible. Similarly, B2 pair (babul bark with Himalayan rhubarb roots) is fairly compatible as per K/S vs. ΔL and ΔC vs. ΔL plots of colour build-up as observed in the conventional method; while the newer relative compatibility rating (RCR) method also shows this B2 pair to be fairly compatible. Lastly, and B3 pair (babul bark with pomegranate rind) is moderately compatible as per K/S vs. ΔL and ΔC vs. ΔL plots of colour build-up as observed in the conventional method; while the newer relative compatibility rating (RCR) method also shows this B2 pair to be averagely compatible. So, the newer and easier method of the RCR technique gives nearly matching results as the conventional method.

Author details

Deepali Singhee[1*], Yamini Dhanania[1] and Ashis Kumar Samanta[2]

1 Department of Textile and Fashion Technology, J.D. Birla Institute, Kolkata, India

2 Department of Jute and Fibre Technology, Institute of Jute Technology, University of Calcutta, Kolkata, India

*Address all correspondence to: deepalisingheejdbi@gmail.com

References

[1] Samanta AK, Agarwal P, Datta S. Studies on colour interaction parameters and colour fastness properties for dyeing of cotton fabrics with binary mixtures of jackfruit wood and other natural dyes. Journal of Natural Fibers. 2009;**6**:27-45

[2] Salimian S, Khalili H, Izadan H, Shahamatjoo S. Feasibility of using digital colour imaging devices for the determination of cationic dyes compatibility. Colour Research and Application. 2017;**42**(3):337-345

[3] Abeta S. Compalibility of acid dyes on nylon. Dyes and Pigments. 1992;**18**:57-65

[4] Datye KV, Mishra S. Compatibility of dye mixtures. Journal of the Society of Dyers and Colourists. 1984;**100**:334-339

[5] Samanta AK, Agarwal P, Datta S. Dyeing of jute fabric with binary mixtures of jackfruit wood and other natural dyes - study on colour performance and dye compatibility. Indian Journal of Fibres and Textiles Research. 2008;**33**:171-180

[6] Shahparvari MR, Mahdi S, Safapour S, Gharanjig K. Compatibility of natural dyes on aluminum pre-mordanted woolen yarns by determination of diffusion coefficient. Fibers and Polymers. 2018;**19**(8):1663-1669

[7] Sultana R, Uddin MZ. Compatibility testing of reactive dyes. Journal of Mechanical Engineering. 2007;**38**(Dec. 2007):61-64

[8] Shukla SR, Dhuri SS. The effect of dye characteristics on their compatibility in mixtures as assessed through colour coordinates. Journal of the Society of Dyers and Colourists. 1993;**109**:402-405

[9] Shukla SR, Dhuri SS. Improving the compatibility of disperse dye mixtures using leveling agents - Assessment through colour coordinates. Journal of the Society of Dyers and Colourists. 1992;**108**:395-399

[10] Blackburn D, Gallahher VC. Disperse dyes for polyester: A new approach to compatibility. Journal of the Society of Dyers and Colourists. 1980;**96**:237-245

[11] Singh M, Bhattacharyya N, Gupte VC. Determination of compatibility of reactive dyes in mixtures by chromaticity diagram. Colourage. 2006;**53**:71-76

[12] Sinnur HD, Verma DK, Samanta AK. Compatibility of binary mixture of natural dyes for developing compound shades for cotton khadi fabric. Indian Journal of Fibre & Textile Research. 2021;**46**(2):158-167

[13] Dar MS, Ikram M. Studies on Quercus infectoria: Isolation of syringic acid and determination of its central depressive activity. Planta Medica. 1979;**35**(2):156-161

[14] Shahid M, Ahmad A, Yusuf M, Khan MI, Khan SA, Manzoor N, et al. Dyeing, fastness and antimicrobial properties of woolen yarns dyed with gallnut (*Quercus infectoria* Oliv.) extract. Dyes and Pigments. 2012;**95**(1):53-61

[15] Vankar PS. Commercial viability of natural dyes: Heena, Harda, catechu and babool for textile dyeing. Natural Product Radiance. 2002;**1**(4):17

[16] Sinnur HD, Samanta AK, Verma DK. Standardization of dyeing process variables for dyeing of cotton khadi fabric with aqueous extract of babul bark (*Acacia nilotica* L.). Journal of

Institution of Engineers India Ser E. 2018;**99**(2):187-207

[17] Rather LJ, Islam S, Shabbir M, Bukhari MN, Shahid M, Khan MA, et al. Ecological dyeing of woolen yarn with Adhatoda vasica natural dye in the presence of bio-mordants as an alternative copartner to metal mordants. Journal of Environmental Chemical Engineering. 2017;**4**(2):3041-3049

[18] Prabhu KH, Bhute A. Plant based natural dyes and mordants: A review. Journal of Natural Product and Plant Resource. 2012;**2**(6):649-664

[19] Bittencourt C, Felicissimo MP, Pireauz JJ, Houssiau L. ToF-SIMS characterization of thermal modifications of bixin from Bixa orellana fruit. Journal of Agricultural and Food Chemistry. 2005;**53**(**16**):6195-6200

[20] Islam S, Rather LJ, Shabbir M, Bukhari N, Khan MA, Mohammad F. First application of mix metallic salt mordant combinations to develop newer shades on wool with *Bixa orellana* natural dye using reflectance spectroscopy. Journal of Natural Fibres. 2017;**15**(3):363-372

[21] Shah BN, Seth AK. Drugs containing glycolides. In: Textbook of Pharmacognosy and Phytochemistry. New Delhi: Reed Elsevier India Private Limited; 2010

[22] Gupta DB, Gulrajani ML. Journal of the Society of Dyers and Colourists. 1996;**112**(8):270

[23] Adeel S, Ali S, Bhatti IA, Zsila F. Dyeing of cotton fabric using pomegranate rind (*Punica granatum*) aqueous extract. Asian Journal of Chemistry. 2009;**21**(5):4671-4680

[24] Satyanarayana DNV, Chandra KR. Dyeing of cotton cloth with natural dye extracted from pomegranate rind and its fastness. International Journal of Engineering Sciences & Research Technology. 2013;**2**(10):2664-2669

[25] Hao L, Wang R, Fang K, Liu J. Ultrasonic effect on the desizing efficiency of a-amylase on starch-sized cotton fabric. Carbohydrate Polymer. 2013;**96**(2):474-480

[26] Dhanania Y, Singhee D, Samanta AK. Optimization of dyeing process variables for eco-friendly dyeing of cotton fabric with babul bark extract as a natural dye and gallnut extract as a bio-mordant. Journal of Natural Fibres. 2021;**19**(13):5478-5495. DOI: 10.1080/15440478./2021.1875955

[27] Evans, Trease WC. Pharmacology. 14th ed. London: WB Saunders Company Ltd; 2000

[28] Harbone JB. Phytochemical Methods - A Guide to Modern Techniques of Plant Analysis. London: Chapman and Hall; 1998

[29] Shah HS, Gandhi RS. Instrumental Colour Measurements and Computer Aided Colour Matching for Textiles. Ahmedabad, India: Mahajan Book Distributers; 1990. pp. 76-116

[30] Fairman HS. Metameric correction using parameric decomposition. Colour Research Application. 1987;**12**(5):261-265

[31] AATCC. Colour fastness. In: AATCC Technical Manual. Vol. 85. U.S.A: AATCC; 2010

[32] Nathan VK, Rani M, Rathinasamy G, Dhiraviam KN. Antioxidant and antimicrobial potential of natural colouring pigment derived from *Bixa orellana* L. seed aril. Proceedings of National Academy of Sciences,

India Section B: Biological Science. 2017;**89**(5):137-143. DOI: 10.1007/s40011-017-0927-z

[33] Rios A, Borsarelli C, Mercadante AZ. Thermal degradation kinetics of bixin in an aqueous model system. Journal of Agricultural and Food Chemistry. 2005;**53**:2307-2311

[34] Agarwal SK, Singh SS, Verma S, Kumar S. Two new anthraquinone derivatives from *Rheum emodi*. Indian Journal of Chemistry. 1999;**38**(B):749-751

[35] Santos J, Monteiro S, Oliveira S, Magalhães P, Magalhães F, Martins JN, et al. Application of forest by-products in the textile industry: Dyeing with pine and eucalyptus bark extracts. Environmental Sciences Proceedings. 2022;**22**:30-36. DOI: 10.3390/IECF2022-13053

[36] Arvindekar AU, Laddha KS. An efficient microwave-assisted extraction of anthraquinones from *Rheum emodi*: Optimisation using RSM, UV and HPLC analysis and antioxidant studies. Industrial Crops and Products. 2016;**83**(May 2016):587-595

[37] Nasr CB, Ayed N, Metche M. Quantitative determination of the polyphenolic content of pomegranate rindrind. Zeitschrift für Lebensmittel-Untersuchung und -Forschung. 1996;**203**:374-378

[38] Qureshi WA, Vivekanandan B, Jayaprasath AJ, Ali D, Alarifi S, Deshmukh K. Antimicrobial activity and characterization of pomegranate rind-based carbon dots. Journal of Nanomaterials. 2021;**2021**:Article ID 9096838. DOI: 10.1155/2021/9096838

Chapter 8

Colorimetric Characterisation and Process Standardisation for Application of Natural Dyes on Textiles: A Research Review

Pubalina Samanta, Adwaita Konar and Asis Muhopadhyay

Abstract

Colorimetric characterisation and to standardise the dyeing processes for applying natural colourants (dyes) on different textiles materials and related investigations have been highlighted here with a broad coverage. Lack of sufficient colorimetric characterisation data for different natural dyes led to the researchers to take a stock of total work done in this field. Without characterisation of colour components and process standardisation of specific natural dye and natural fibre combination, reproducibility, uniformity, maximum or optimised colour yield and acceptable grade of colour fastness could not be possible to achieve for natural dyeing of different textile materials. Hence, a systematic review of standardisation of extraction conditions of natural dyes and finishes, type of mordant and mordant application conditions and finally standardised dyeing process variables have been elaborated. Worldwide eco concern of consumers have created scope of newer research for revival of natural dyes and natural finish application on textiles with standardised methods/process conditions to obtain reproducible dyeing results with acceptable colour fastness. Functional finishing of textiles for antimicrobial and UV resistant finishes by natural resource materials are the two relatively newer areas, needing further research. In the concluding remarks advances in natural dyeing and natural finishing have also been dealt with futuristic need to explore.

Keywords: colorimetric characterisation of natural colourants and dyed textiles, extraction of natural colourants, natural mordants, standardisation of natural dyeing process, identification of natural dyes and pigments and natural finishes

1. Introduction

Natural colourants (dyes) are extracted from various natural bio-logical resources. The origin may be any parts of vegetable plant or animal. Although, there are exceptions where the source of colouring matters is of mineral origin. 'The Society of Dyers and Colourists' (SDC) in their 3rd edition specified 'Colour Index" as natural colouring/dyeing substances and defined as: "The natural dyes and pigments comprise of all the colours obtained from vegetable and animal matters with no or very little chemical treatments". These natural colourants are mainly anionic type dyes having mordant-able phenolic-hydroxy or carboxylic groups rendering necessary coordinating capacity with metallic and other type of mordants, with few exception having either cationic nature (barberry) or class of insoluble vat dye type (natural indigo), where they can be classified under application based classification of textile colourants such as anionic direct dyes (turmeric), as cationic basic dyes (barberry), as anionic acid dyes (madder/Manjishtha) with or without mordanting ability, also as non-ionic insoluble vat dye type (natural indigo) and few disperse dye/solvent dyes type (Henna/Lawson) besides few are also available as pigment colour like Azurite/Ash/ Limonite [1–6]. Majority of the natural dyes are non-substantive for most of the textile fibres, which means that they have little attraction towards any fibres i.e., they have no colouring power to dye textiles by applying only themselves, and hence require the aid of some anchoring agent called "mordant". These mordants along with other additives used for dyeing of textiles materials with natural colourants require to be environmentally-friendly as well. Some of the natural substances have medicative values too and may incorporate either microbicidal or UV radiation protection or both, which need to be explored for natural finishing of textiles along with natural dyeing.

The production process of synthetic chemicals is associated with series of chemical reactions which consumes high energy. In addition, many unwanted bio-hazardous by-products generate during the chemical reactions [6–9]. These toxic or bio-hazardous by-products are released into different natural streams of water, lakes or in the biosphere. These negative impacts of the man-made synthetic colourants have encouraged researcher to find alternative environmentally-friendly dyes. Now a days consumers are very aware about the ecology and demanding bio-friendly natural products as well as eco-friendly processes like dyeing of natural fibres with natural dyes, which has thus become indispensable. However, all natural colourants may not be necessarily eco-friendly. Hence, ecological hazards associated with any natural colourants should always be examined/asserted before use. The production of man-made dyes involves series of hazardous chemical reactions and produces lot of effluents and objectionable gases causing water and air pollution and are usually conducted at high temperature and pressure causing lot of energy consumption and is carbon negative process, using so many primary chemicals isolated from petroleum derivatives [10–13], while natural dyes and pigments have manifold advantages over synthetic dyes and pigments, as they are eco-friendly without any toxicity and safe for human. Majority of the natural colourants are evidenced to be ecologically safe without toxicity except few. Hence, although synthetic dyes show better performance, people are more attracted for application of natural colours for textiles due to growing knowledges that most of the natural colours have neither allergic consequences nor toxic effects. In addition, textile products dyed with natural colouring matters are comfortable for the wearer with soothing colouring effect [14–20].

However, the present scenario there are an estimated around 30 million tonnes of consumption of textiles across the global and such mammoth quantity of textile products may not be coloured with only natural colourants [19]. Hence, the use of ecologically green synthetic dye-stuffs cannot be avoided or fully replaced by natural dyes, but a small part of dyed textiles may be handled by bio-friendly natural colourants, ensuring green measure for colouring for a portion. Application of natural colourants will also render new avenue for engaging rural population because plants of natural colourants are normally grown at country side at the barren lands. So, those waste lands can be used to grow the respective dye producing plants or herbs [19]. However, for reproducibility, acceptable colour fastness and uniform dyeing in either single or compound shade using combination of two or more combination of natural dyes, standardisation of dyeing input variables and analysis of compatibility for use of blend of these natural colours are crucial [21, 22].

Today, when the world is endangered with environmental pollution-oriented destruction from the man-made toxic chemicals (including hazardous synthetic dyes and its intermediates), awareness about sustainability should be the main focus, otherwise, it would be impossible to save our planet earth from annihilation. That is why it is important to revive the ancient art of colouring with "*Natural Colourants and Pigments* " [10, 21, 22]. Some other associated advantages of natural dyes and pigments include expected non-toxicity and anti-allergic properties and some medicinal value [23] as well as anti-microbial property and UV protective character [24]. But, many times natural dyes and pigments lack in uniformity and reproducibility of shade. Limited availability of such natural dyes and pigments is major drawback in addition to availability of handful scientific information on mechanism [25] and standardise dyeing process controlling variables [21, 26] for deciding proper methods of dyeing with natural dyes and pigments and colour fastness [27] achievable for such natural dyed textiles, which are based on chemistry of such natural dyes and pigments and their possible interaction with mordants and textile fibres on which those are applied. For Improving dyeing yield and dyeing performances, for dyeing textiles with natural dyes, many researchers attempted either modification of textile fibres to improve colour yield on modified textile fibres or attempted to adopt newer methods/techniques of dyeing textile fibres with those natural dyes [28–33]. To assure that a textile substrate is dyed purely with natural dye, there is need of developing suitable methods and standards for identification [34] of natural dyes and pigments identifiable from such dyed textiles for consumers' assurance and satisfaction. Use of extract of neem bark/leaf [35] or chitosan [36] derivative and extract of tulsi leaf [37] have been referred to have antimicrobial character and extract of anar-peel [38] has been referred to have UV protective character besides their colouration effect on textiles except chitosan.

2. Principles of application of natural dyes on textiles

'The Society of Dyers and Colourists' (SDC) in their 3rd edition specified natural colouring substances as '*The natural dyes and pigments comprise of all the colours obtained from vegetable and animal matters inclusive few vat, disperse/solvent dyes, pigment colour and a few direct, basic and acid dyes*' [1, 2, 19].

Majority of natural dyes are mostly non-substantive towards textiles, and hence, it is normally be used along with a mordant or so called mordanting assistant compounds. A mordant usually a metallic salt or other coordinating complex forming agents having an attraction towards both the colouring component and the fibre and thus able to form complexes combining the fibre-mordant-natural dye, forming an insoluble complex as precipitate or lake of complex compounds remain anchored on the specific textile fibres. Application wise, natural dyes include mostly either direct dyes or acid dyes or other classes of mordantable dye type. There are some natural dyes akin to vat dyes, a few are as solvent dyes and also as pigments also. Only one natural basic dye and one or two disperse dye type natural dyes are known, but natural dyes akin to sulphur dyes, azo dyes are still unavailable.

Natural colourants are usually non-substantive, needing metallic or other types of mordants, or other complexing agents as so-called mordanting assistant, requiring for attracting and fixing natural dyes to the fibre substrate. Hence natural dyeing of textiles has the following three important stages for successfully applying it on textile fibres.

2.1 Extraction process of natural dyes from source materials

Extraction process of natural dyes from source materials by applying aqueous method i.e., extraction with the help of boiling/hot water with or without added any acidic/alkaline/alcohol containing chemicals in extraction bath within fixed ML ratio, time, pH and temperature range followed by distillation, dehydration and vacuum drying etc. or by supercritical carbon dioxide fluid extraction process or solvent extraction using soxhlet apparatus extraction by mixture of alcohol and benzene or by rotatory vacuum pump or extraction under low pressure etc. [1, 2, 39].

2.2 Use of mordant/mordanting assistants for creating substantivity and fixation of natural dyes to the fibres

The origin of word 'mordant' is from '*modere*' (Latin) meaning '*to bite*'. The mordant attaches with the fibre surface so that a dye can sink in it. It can be tannic acids or tannins, sulphonated oils, metal salts or any natural product containing tannic acid/tannates.

These mordants not only fix the natural dyes on fibre but also govern their colour produced on the fibres and thus provide a variety of colours with a same dye material by using different mordants. Attached mordants play an important role in fastness and intensity of colour properties during the subsequent dyeing operations. Mordants thus fulfil following functions as described below:

1. Dye fixing on the fibre,

2. Increase the fastness properties of dye,

3. They themselves may act as a colouring matter,

4. They may ease dyeing/enhance dye absorption.

The followings chemicals/extracts of natural agents may be used as mordanting assistants:

a. *Tartaric acid*: help to acidify dye bath solutions and can used to nullify calcareous lustre effect.

b. *Tannic acid and chebulinic acid*: the natural 'tanning agent' seems usual in natural dye fixation as mordanting assistant. A number of tannins containing substances are employed as mordants or mordanting assistant during natural dyeing process. Among these raw tannins, myrobolan (*harda*) containing chebulinic acid, gall nut and Sumach containing tannic acid are most important.

c. *Tartar cream:* it is a whitish crystalline material normally utilised for colligation with mordants to brighten the colours.

d. *Glauber's salt/common salt*: it acts as levelling agent used extensively in silk dyeing.

e. *Boiled-off liquor*: during use of degumming solution of silk in the dye-bath of silk, it helps the colouring substance to be distributed evenly and it also helps it to preserve lustre.

f. *Enzymes/alkali:* use of compatible alkali/enzymes combination in pre-treatment processes or into the dye bath containing natural colourants, improve colour yield.

2.3 Dyeing methods and principle/mechanism of fixation of natural dyes on textiles

Natural colourants frequently display non-consistent shades due to varying dyeing parameters or not adopting standard dyeing conditions. So, standardisation/optimisation of colouring/dyeing process for particular dye-fibre-mordant co-relations are necessary. Being mordantable, dyeing of majority of the natural dyes is accomplished by aqueous extracts of dyes in normal/hot dye-bath. Colouring may be executed by pre-mordanting, post-mordanting or concurrent mordanting, followed by standard dyeing process in normal/hot aqueous bath or HTHP technique with controlled pH range (i.e., acidic or alkaline), or using advanced ultrasonic dyeing technique etc. [1, 2, 39], using single or more natural colours extracted appropriately and found compatible scientifically.

Among different metallic salts most common eco-friendly mordant largely used in natural dyeing is alum (potash alum and ferric alum) besides uses of ferrous sulphate, stannous chloride etc. with or without harda/myrobolan as mordanting assistant, two types of natural potash alum are available in the market in two forms—potash alum containing 10.8% of $Al_2(SO_4)_3$ and ammonia alum containing 11.9% of $Al_2(SO_4)_3$.

Usually only alum or 'phatkiri' means potassium-alum, which is a hydrated complex of potassium-aluminium-sulphate with formula $KAl(SO_4)_2{\cdot}12H_2O$. In cases of mordant-able natural dyes, aluminium sulphate or other metals chemically combine with some —OH or —COOH or other specific groups present in the dye and then further bound by co-ordinated/covalent bonds or hydrogen bonds and other interactive forces with the accessible functional groups present in the fibre [2] to form fibre-mordant-dye complex (for fixation of natural dyes on fibre substrate).

2.4 Advantages and limitation of natural dyes

In present days, there is an increase in demand of natural colourants. Although synthetic dyes have superior fastness properties, good durability and relatively easy to apply, the natural colours have certain limitations over synthetic dyes. Some of advantages [1, 2, 39] are as follows:

a. Moderately non-polluting and less toxic.

b. Soft, lustrous eye soothing shades to eye.

c. Vast range of colours. A little variation in dyeing conditions or by using other mordants for a particular dye can produce a new colour, which is impossible with man-made dyes.

d. Natural dyestuffs produce uncommon colour ideas with harmony.

e. Natural dyes are biodegradable as well as replaceable with saving of energy.

f. In few cases, like harda, indigo etc., the wastage can be utilised as fertiliser for agricultural fields.

g. As many plants/herbs of natural dye are grown on wastelands, utilisation of wasteland is an additional advantage for the natural colourants. Dyes like madder developed as host in tea gardens. So, no extra cost required for its cultivation.

h. Being highly labour-intensive, it can create new job opportunities like cultivation, extraction of natural colours and their application.

i. It has a potential to uplift the textile export of the country and thus may increase earning of foreign exchange [2].

j. Naturals dyes are safe for skin being anti-allergens and non-hazardous.

k. Natural dyes are good moth resistant.

l. Some natural dyes become more attractive with ageing, while synthetic dyes fade away with time [1, 2, 18, 26].

m. During bleeding, for any reason, natural dyes do not stain the adjacent fabric except turmeric.

n. Natural dyes can good UV protector without altering wear properties.

o. They are utilised in repairing and preservation of historic textiles materials.

p. They can be an ideal replacement synthetic dyes for food-stuffs without any hazards.

Despite these advantages, natural dyes possess some in-built disadvantages [1, 2, 39], such as:

a. Due to non-availability of sufficient database and standardising recipe for different shades on different textiles, there is doubtfulness about final shades and different fastness properties to be achieved and the same variation from plant from different sources, between crops and different seasons (i.e., variation due to different location, different crops and different seasons) and even for different mordant.

b. The extraction of dyes, preparation of textile, mordanting need competency which is expensive. Application of costly mordants, use more dyestuffs with long dyeing time to compensate low colour yield raise the dyeing cost for natural colouring as compare to man-made dyes but it has a niche market for bio-friendly, sustainable products dyed with natural colours particularly expensive products of silk, jute, wool, cotton and even for nylon etc.

c. Insufficient knowledge base due to less exploration it needs more studies/ researches for commercialization.

d. Limited knowledge available on dye wise extraction process and fibre-wise dyeing techniques for natural colours cause uncertainty for shade depth and various colour fastness characters.

e. Natural dyes have a unique characteristic to change its colour during exposure to sun light (UV ray), perspiration, moisture and air [39] which may be considered disadvantageous.

f. Nearly all-natural dyes require to use of metallic mordants as dye fixing agent to fabric. During colouring, a prominent part of the metallic mordant remains in the residual dye bath unexhausted and these metals are threat to environment. So, hazardous metallic mordants containing metal like copper/chromium etc. should to be eliminated.

g. Being fugitive nature, these natural dyes are applied in conjugation with a mordant causing inadequate fastness to colour which is mis-match with textile end usage. Acidic pH during dyeing or post treatment processes for natural colours to enhance fastness to colour is hazardous for biosphere.

h. Few natural dyes have a little bit toxicity but there are limited studies available to assess their toxicity. There are only one or two reports available on source of natural dyes/colour which have been figured out as a potential threat to ecological balance due to toxicity, e.g., quercetin is studied to be mutagenic.

i. There are very few trained professionals available for application of natural dyes. So, there is a need for in-depth training and interaction programme for successful application of natural dyes.

j. Non-availability of pure natural dyes in the form of powder or in liquid form for direct application for developing known shade depth.

In spite of these limitations associated with application of natural dyes, it has a niche export market for dyed/printed fibres provided complying with the GOTS standards originated from the developed countries combining perniciousness, eco-friendliness and pollutant norms. For revival [40–46] of application of natural dyes these disadvantages are to be partially or fully eliminated/reduced utilising colorimetric analysis of sufficient scientific research data.

2.5 Colour fastness for natural dyed textiles

Fastness to colour is the resistance of colour component to alter its any colour features or to shift its colourants to neighbouring materials or both. Fading indicates the alteration of depth of colour (lightening or darkening) during laundering or friction during rubbing or during human perspiration or on exposing to light. Some reports in literature [47–50] has described effective ways and means for improving colour fastness to wash and light by using different pre or post treatment with natural or eco-friendly fixing agents including tannin/chitin or quaternary compounds or UV absorbers related natural or substituted eco-friendly materials. There are many known techniques available to improve wash-stability, fastness to rubbing and light for synthetic dyed products, but such techniques for natural colours are still very limited. So, exploring fastness improving techniques for natural colourants are essentially needed as compared to synthetics.

Fading during exposure to UV light is generally boosted up in presence of vapour, thermal energy, oxygen in air and also depend on many other factors, which are shown in **Table 1**. Light fastness property of natural colours depends on its UV-soaking property of respective dyes [49, 50]. For example, UV absorption character of pomegranate rind i.e., anar peel extract (*Punica granatum* L.) has been proven and highlighted by many researchers in past [51–55], which caused improved light fastness and higher Sun protection factor for pomegranate rind extract dyed natural textiles.

Poor fastness to washing for majority of the natural colours is mainly associated with weak dye-mordant-fibre bonds. Breakages of dye-(metallic) mordant-fibre compound during washing by ionisation of the natural colours is responsible for change in hue [12]. Due to ionisation of many phenolic hydroxy groups present in natural dyes under acidic or alkaline condition, fabrics coloured with natural colourants alter its colour characteristics during laundering with alkaline surfactants. In most of the cases weak hydrogen bonds is responsible for poor washing fastness. In such cases, dyeing isotherm shown may be Nernst isotherm type showing high value of partition coefficient, like absorption of disperse dye on polyester fibre. Factors on which light fastness of any natural or synthetic dye depend are summarised in **Table 1**.

Rubbing fastness is measured by change in colour on the surface of rubbed textile or staining to the abrader cloth in both dry and wet conditions (for wet rubbing fastness, bleeding of colour may occur due to migration) after completion of specific

Factors	Examples with and comments
Energy level and exposure time to light	Strictly regulated by light fastness test methods (Sun-light, Xenon arc, MBTF lamp)
Dye-stuff or pigment: type, density and dispersion	Superior light fastness with higher concentration and low specific surface area (pigment shows good light fastness than molecular dispersed dyes). Matching dye distribution over the fibre cross-section shows superior light fastness than ring dyeing
Dyestuff combination i.e., mixture of dyes	Catalytic fading (one dye catalyses photo-degradation of another dye), rarely it gives light fastness protection sometimes
Fibre type	Some dye gives poor light fastness to jute than onto cotton or acrylics due to self-fading behaviour of the fibre itself for jute
Moisture	Better is light fastness with lower moisture content. However, the level of humidity is usually fixed for standard methods of light fastness testing
Heat	Less is the light fastness rating with higher temperature; however, temperature is pre-fixed by standard light fastness testing method

Table 1.
Factors on which light fastness of natural dyed or synthetic dyed textiles depends.

abrasion cycles. Depending on the tensile strength of the fibre, small coloured particles may bleed out showing poor fastness to rubbing. If the dyes are not affixed properly on the fibre surface, that will stain the abrader fabric during rubbing. Numerical rating for rubbing fastness ranges from 1 (very poor) to 5 (best) depend on the following parameters:

1. Molecular structure and size, chemical characteristics and mode of fixation of natural dyes,

2. Mordants type and concentrations

3. Concentration, shade percentage and diffusion of dyes

4. Dyeing process, fixation percentage and degree of surface dyeing and ring dyeing

5. Amount of unfixed surface deposited dyes not removed after washing

6. Application of dye fixing chemicals, softener, silicones and cross-linking agents.

3. Colorimetric evaluation and process standardisation of natural colouration for textiles

There are various difficulties associated with natural colourants due to insufficient knowledge and lack of scientific information to attain uniform and optimum colour yield with acceptable colour fastness characteristics for silk fabrics. No such integrated study is available with pomegranate rind (anar peel) applied on silk, except one recent study on standardisation of dyeing process variables using pomegranate rind on cotton fabric [55].

Unknowingly natural dyes had been in use form primitive age of civilization in different regions of the world. It has been used since pre-historic times in order to beautify various materials like foods, drinks, household articles, fabrics etc. Almost all the dyes were of natural origin till 1800 B.C. As mentioned in literature [56], William Perkin synthesised first man-made dye 'Mauviene' by in 1856. Some studies on the use of synthetic dyes had reported that these dyes are not only non-biodegradable but also pose a threat to human health in terms of causing irritation, genotoxicity and even carcinogenicity. With the present sustainable concept to use of environmentally friendly and bio-degradable substances, the application of natural dyes on natural textiles have gained the momentum again and awareness of consumers about eco-friendly textile merchandise, use of natural colours is slowly reviving again [40–46]. There are several advantages of natural dyes over synthetic colourant/dyes [1, 2, 12, 39, 57] and such eco-friendly, nontoxic natural dyes are safe even for body contact [58]. Many scientists had already reported about the medicinal value of many natural dyes [51, 59–60]. Yellow natural colourants extracted from rhizome of turmeric was traditionally utilised for medication in case of anti-inflammation [61].

The natural colours are extracted from the vegetables/plants/herbs, minerals or insects [62]. Majority of the natural colours show inferior to average light fastness property whereas synthetic dyes exhibit moderate to excellent light fastness property

[63]. Dyeing of cellulosic fibre with leaf-extract of *Beilschmiedia fagifolia* had been reported in literature [64], where sonicator assisted dyeing method, that was used to dye cellulosic fibre with *B. fagifolia* extracts. The researchers had observed that cotton pre-treated with 1–2% metallic mordant and dyed with 5% of plant extract developed good fastness by use of sonicator assisted dyeing process. In another study [65] marigold flower dye extract was utilised to dye cotton fabrics.

Natural dyes leaching out from *Jatropha integarrima* flowers was used for colouring of different textiles made of cotton, wool, silk [66]. Silk dyeing with *Onosma echiodes* (Golden drop) had also reported in literature [67]. Areca nut palm extract was used as natural colourant [68] to dye cotton with suitable mordant. Colouring of natural cellulosic material and silk with advance ultrasonic technology by *Nerium oleander* flower petal extract had already been studied [69]. Colour extracted from waste leaves of *Aroto carpus* betero phyllus has been reported to dye various textile materials made of cotton and silk to obtain a consistent golden yellow shades [70]. Successful colouring with natural colourants extracted from turmeric, catechu, henna, madder, tea leaves, Indian rhubarb and pomegranate exocarp on man-made polyamide fibre had been reported in literature with details of procedure and dyeing performances [71]. Some successful researches had also been carried out on application of Lac dyes and its absorption criteria [72] on different fibres. Mulberry silk was successfully dyed with natural dye annatto [73]. Some studies [74–76] on application of natural dye of Indian madder, *Spathodeac ampanulata* and lac dyes on silk had been reported. Cost effectiveness of dyeing silk in different shades with different natural dyes was analysed and reported by Patel et al. [77].

The substantial scientific document on natural colouring matter was initiated by Perkin and Everest [78]. Studies on contemporary processes for applying natural dyes on textiles as practiced in different states of India along with the chemistry of such dyes and traditional dyeing methods has been well documented by Mohanty et al. [13]. This book is probably an important authenticated document on the processes of natural dyeing on textiles of ancient India till today. The present section of literature review thus covers some important research work on application of natural colouring matters on textiles under the following sub-heads:

1. *Characteristic and bio-chemical/chemical anatomy of some important natural dyes.*

2. *Methods of extraction of colourant component from natural dyes.*

3. *Effects and methods of application of different mordants.*

4. *Different methods of natural dyeing and variation of dyeing parameters.*

5. *Study of dyeing kinetics and dyeing process variables.*

6. *Exploring new shades with natural dyes and colour parameters analysis for dyed textiles.*

7. *Colour fastness properties of natural dyes.*

8. *Miscellaneous recent studies.*

3.1 Characteristic and bio-chemical/chemical anatomy of some important natural dyes

Studies on analysis of natural colours and their application on textiles are a challenging subject which attracted of scientist and academicians for compiling it in a form of scientific literature and books as early as in 1918 since the publication of book by Perkin and Everest [78]. Since then, among all review and books published on the chemistry of natural colours and pigments and their application on textiles. Some ancient and present literature reports are specially mention worthy as reported by Gulrajani and Gupta [1], Samanta and Konar [2, 10], SERI [12], Mohanty et al. [13], Vankar [39] and others like Perkin and Everest [78], Sahid and Mohammad [79] and Mayer and Cook [80], to utilise as study materials for further reading.

Structures of quinoids based natural colourants is described in Thomson's book [81]. Earlier review work of many authors on natural dyes and pigments also include the review report by both Parris [82] and by Hofenk and Graaff [83], where the latter collected data about fastness to colour and history of their uses. A detailed review work was done by Samanta and Agarwal [4], Samanta and Konar [2, 10], and in a recent book of Vankar [39].

Samanta et al. [84–86], had analysed the thermodynamics of rate of dyeing, half dyeing time, internal energy of system (i.e. enthalpy), dyeing isotherms, free energy etc. as a physico-chemical attributes of colouring for ligno-cellulosic material with natural dyes like red-sandal wood, jackfruit wood and tesu (palash) etc.

Mass spectroscopic analysis of dyeing of cellulose polymeric material with indigo has already studied and finger print region of it was used for identification of natural indigo [87] with the supporting analysis of TLC and UV VIS spectroscopic analysis. However, there are contradictory reports on non-iso-labelling/non-identifying natural indigotin differentiating it from synthetic indigo as reported by De Wong [88], who was unable to identify 6,6-dibromo-indigotin (with the help of direct ingredient index analysis), but later it was resolved after the 6,6-dibromo-indigotin had been separated from natural indigo by subtractive extraction process with sodium hydro-sulphite. Koren [89] had analysed the natural madder and natural indigoid dyes by high performance liquid chromatography (HPLC) to figure out its chemical functional nature.

Thin layer chromatography (TLC) analysis is very useful tool for researchers for identification of natural colourants in textiles and Parris reported TLC [82] different vegetable and insect dyes of yellow, red and blue shade. Guinot and Roge [90] also used TLC as simple chromatographic analysis technique to study extracts of different parts of herbs/plants bearing flavonoids (flavonols, flavones, flavanones, chalcones/aurones, anthocynanins), hydroxy-cinnamic acids, tannins and anthraquinones, which are the active phyto-colouring substances found in those plants. The colour elements were separated from most of the barks containing bio-flavonoid moieties.

Physico-chemical dyeing attributes of natural colours extracted from red sandal wood and its compatibility behaviour with other colours were also examined and analysed by Samanta et al. [91, 92]. Analysis of UV VIS spectroscopy of extract from the bark of neem tree [93] extracted colourant showed dual maximum absorption bands at 275 and 374 nm respectively; while extract of beet sugar exhibited three absorption bands at 220, 280 and 530 nm. The visible spectrum of extract of ratanjot was studied by Gulrajani et al. [94] under acidic medium and it displayed highest absorption around 520–525 nm, while under alkaline pH absorption band was shifted to 570 nm with an additional peak at 610–615 nm. Extract of Red sandalwood [94] showed a substantial absorption crest at 288 nm but highest absorption was observed at 504 nm and 474 nm at pH 10 in methanol extracted solution. Colourant extracted from *Gomphrena globosa* flower showed single major peak at 533 nm and there were almost no changes observed in visible spectra at pH 4 and 7; however, there was little shift of peak to 554 nm as observed by Shanker and Vankar [95].

Extraction, identification and assessment of dyeing behaviour on wool of natural dyes obtained from barks of *Musops elengi* and *Terminilia arjuna* were studied by Bhuyan et al. [96] and they observed the dye absorption by wool fibres varied for respective dyes extracted from *Mimusops elengi* and *Terminalia arjun* depending on their dye contents. The dye absorbed by wool fibre varied in between 21.94–27.46% and 5.18–10.78% respectively for the said two natural dyes.

Analysis of colorimetric parameters including *K/S* values, DE values, L, a, b, DC, DH, DL values etc. was studied for use of binary dye mixture for obtaining compound shades along with analysis of compatibility in between different pairs of natural colour combination applicable for both conventional methods and a newer method for compatibility test of natural dyes based on analysis colour difference index (CDI values), a newly defined and established index of colour differences parameters [97, 98] as a new indicator called 'Colour Difference Index' (CDI), calculated by an empirical formula presented by them. By this newer route, the differences of CDI values for dyeing with binary mixture using combination of any two dyes in different proportions has made it possible to determine dye compatibility rating from the chart given by them, as much simpler and easier.

Identifying the colourants in historical textiles materials by the help of modern spectrophotometric and chromatographic techniques and by sensitive colour responses were studied by Blanc et al. [99], who also studied the retentivity behaviour of natural colour components like carminic acid, indigotin, corcetin, gambogic acid, alizarin flavanoid, anthraquinone and purpurin after dyeing. There is another non-destructive way via study of emission and excitation spectra, which was reported for recognising faded dyes on textiles fabrics. Zin et al. [100] characterised purified extract of natural dye components extracted from bark of mango tree for application in wool fibres.

Walker and Needles [101] studied the method for segregation and identification of natural colours from natural protein fibre like wool by reverse phase HPLC using a C-18 column. Binary and quaternary solvent systems were utilised to find chromatograms of dyes, isomers and minor products present in the dyed sample. A linear gradient elution method was used in HPLC analysis for natural colour extracted from insect, plant/herbs/trees, for red anthroquinonoid mordantable colours, red purple natural colour, molluscan blue and indigoid vat dyes [89]. This technique was very useful for identifying various chemical categories of natural colourants from the same elution programme. In addition, it is much time saver for natural anthroquinonoid dyes over those previously published.

Cristea and Villemar [50] reported successful quantifiable analysis of Weld by HPLC and concluded that 0.448% luteolin, 0.357% luteolin 7-glucoside and 0.233% luteolin-3′7-diglucoside in aqueous methanol solution can be after identified within 15 minutes. Some other authors [102] reported analyses of indigo by HPLC that with increase in dyeing time, it changes structure of indigo compound causing a reduction in colour intensity of polyester fibre-based textiles. Jain and Vashanta [103] analysed antimicrobial nature of bio-friendly natural colourant with arcea nut with natural mordanting add-on like myrobolan, lodhra and pomogrenate rind and observed that pomegranate rind showed optimum anti-bacterial performance and Lodhra renders best washing fastness to among all the tannin based natural mordants/mordanting assistant/additives used.

Mondhe and Rao [104] had attempted to develop azo-alkyd natural colours by reducing nitro-alkyds, followed by diazotization of amino-alkyds and coupling with various phenolic substances present in situ in the seed oil of *Jatropha curcas* and analysed it functional nature by IR spectra.

Majority of the natural colours are non-toxic and eco-organo-friendly. The toxicity analysis [105, 106] for each of natural dyes is therefore necessary and it may also provide positive or negative evidences about the skin friendliness or its adverse effect to human body for any such natural dyed textiles. Few natural dyes and dyed products may also contain toxic component in its composition not skin friendly to human. Primary concerns of toxicity are the human skin toxicity, irritation on skin or eye and sensitization potentiality. In addition, possible long-lasting effects such carcinogenic, reproductive toxicity or mutagenic toxicity had also been identified from many natural products. The LD50 test is popular for its perniciousness. It is said the 'lethal dose for 50% of the test animals' which is the amount of material in kg/kg of body mass, can kills 50% of the animals. Another report describes the toxicity and antimicrobial test data of raw methanolic extracts from different parts of the plant *Artocarpus Hetrophyllus* including stem, roots, leaves, fruit, seeds [107] and their subsequent partitioning with organic solvents like petrol, dichloromethane, ethyl acetate and butanol fractions, rather exhibited a wide spectrum of anti-bacterial performance. The butanol fractions of root/bark and fruit were also witnessed to be good antimicrobial nature but no anti-fungi property.

Tannins was extracted from gall nut and leaf from oak plant (i.e., Oak leaf) by Mishra and Patni [108] from the different plant at Himalayan region. These extracts from gall nut and Oak leaves are rich of gallic acid and tannic acid for better dye fixation for application in colouring of cotton, woollen and silk textiles in presence of various metallic mordants to obtain better colour fastness without any dermatological problems. Jain [109] has shown supremacy of few natural mordants for colouring of cellulosic fabrics with extract of jamun tree and its by-products, inspiring further study on such bio-mordanting and bio-additives on cotton, jute, wool and silk textiles.

Benencia and Courreges [110] studied chemo preventative aspect of red sandalwood oil on papilloma skin of mice and also studied its preventive nature against development of skin tumour for CD-1 mice and anti-viral function against herpes simplex virus-1 and 2. Singh et al. [111] examined anti-microbial activities against pathogens like *Escherichia coli*, *Bacillus subtilis*, *Klebsiella pneumoniae*, *Proteus vulgaris* and Pseudomonas by *Acacia catechu*, *Kerria lacca*, *Quercus infectoria*, *Rubia cordifolia* and *Rumex maritimus*. Minimum repressive/inhibitory density was observed varying from 5 to 40 μg per kg of textile materials. Analysis of natural components of colours for dyed historical textiles was also reported by Ferreria et al. [112] by different methods. Another analytical report from Hauluk et al. [113] render analytical data for ellagic acid content in European oak wood for its use as natural mordanting additives as dye fixative in natural dyeing of different textiles. A realistic evaluation of dyeing of different textiles with vegetable colourants with metallic mordants need to be subjected for determining their toxicity for objectionable metal present in the metallic mordant, which was reported for use of different metallic mordants during natural dyeing process [114].

3.2 Methods of extraction of colourant component from natural dyes

It is reported in literature by Dayal and Dobhal [115] that natural colourant extracted from eucalyptus leaves, *Cassia tora* seeds and bark of *Grewiaoptiva*, shorearobusta by using aqueous medium under different conditions leads to variety of attractive light shades on cotton. Later, Vankar and another group of researchers [28] used the supercritical carbon-dioxide fluids to taking out/extraction and purification of natural dyes from bark of eucalyptus. Extraction of natural dyes/colourants from bio-mass products of cutch, ratanjot and madder were described by Khan et al. [107]. Ray Maulik et al. [116] extracted natural dyes/pigments from Hinjal, Himalayan rhubarb and Jujube bark for dyeing of wool. Teli et al. [71] extracted natural colourants from the coffee-seed for dyeing of nylon. Pan et al. [117] applied natural colours on ligno-cellulosic (Jute) fabric with extraction of deodar leaf, eucalyptus leaf and wood of jackfruit by soaking and boiling it for approximately 4 hrs separately in soft water. Verma et al. [118] attempted to develop dye powder from wattle bark by soaking it overnight in deionised water, boiled in pressure vessel followed by passing through filter to get 15–20% (w/w) dye residual powder. Saxena et al. [119] initially extracted antioxidant components of natural colourants from leaves and fruits of confers cultivated in Iran and also from marigold and chrysanthemum flower petals by simply boiling dried petals in acidic aqueous solution. While, Sarkar et al. [120] extracted natural colour from marigold flower and applied on cotton, wool and silk. Later Sarkar et al. [121] again made solvent assisted extraction of yellow vegetable colouring matter including extraction of marigold flower for its application on even hydrophobic textile substrate. Deo and Paul [122] extracted natural colour/dye components from onion skin by adding of 5% solution of sodium chloride in a ratio of 1:2 and applied with natural mordant combination for dyeing ecru denim fabrics. Dixit and Jahan [123] extracted euphorbia continifolia leaves under acidic pH to dye silk fabric and later evaluated its various colour fastness properties.

Sudhakar et al. [124] derived natural colours from nuts of *Areca catechu,* which is profusely grown in India and can be used for colouration of natural protein fibre like silk or wool. It was reported by Radhika and Jacob [125] that natural colour/dye extracted from jatropa seeds under alkaline conditions, which when is applied on cotton fabric, it gives an attractive range of soft, bright and uniform shade on cotton fabric. Extraction of natural colours from this seed waste/food waste can be achieved better with ethanolic or acidified (with 40 M oxalic acid) extraction. Hagerman and Wilson [126] has described a method for quantitative estimation of ellagic acid from tannin containing natural colouring agents for precise estimation of the same, which are believed to be have good property of natural dye fixation on textile substrate for presence of tannate/ellagic acid component along with colour component(s).

Extraction dye from the teak leaves in methanolic solution was successfully carried out by Nanda et al. [127] to develop a brick red shade for silk/wool dyeing with different mordants. Bhattacharya et al. [128] studied the effects of mordants for dyeing polyamide fabrics with some eco-friendly natural colourants and attempted to standardise natural dyes extracted from arjun bark, babul bark and pomegranate rind for reducing the effect of dyeing process variability. Onal [129] attempted to dye wool, cotton and leather with onion skin/peel (*Allium cepa* L.) extract as natural colourant source.

Agarwal et al. [130] studied superior dyeing properties on polyester by organic solvent (solvent:water ratio 1:9) assisted dyeing using the said solvent dyeing medium, where colours of the selected natural dyes was extracted from henna leaves, and directly used in dyeing of polyester in the solvent dye-bath. Houlton et al. [131] studied the dyeing of jute-cotton blends by extracts using grape skin waste.

Teli et al. [132] had dyed cotton with concentrated tea extract after different pre-treatment of cotton. It was studied by Padmaja et al. [133] to extract natural colourant from Hibiscus flower to apply it on pre-mordanted silk textiles.

Thus, the role of solvent/acid/alkali in both extraction and dyeing with selective natural dyes is understood from above few case studies of earlier work. Hence it is felt appropriate to study the extraction of natural dyes or natural finishes under solvent based/acid based or alkali based extraction, for dissolving out or taking out colour components and other active functionality species for extracting colours from specific natural colourants. Extraction of natural finish components and colour components from extract of catechu and gall nut and eucalyptus leaves need to be studied in depth.

3.3 Effects of different methods of mordanting and applying different mordanting assistants for natural dyeing

Mordanting with selective metal salts can be accomplished by pre, simultaneously and post-mordanting process. Various metallic and non-metallic bio-mordants may be used on different textiles to enhance colour absorption and dye uptake of natural dye. Without mordants, natural colourants have generally no affinity for textile fibres. Extensive work on type of mordants and mordanting conditions including application of few bio-mordants before applying natural colourants/bio-colours on several textiles have been studied in few current literature [134, 135].

Dayal and Dobhal [115] studied the effect of copper sulphate ($CuSO_4$) and potassium di-chromate ($K_2Cr_2O_7$) on natural protein fibre like silk, wool and also for cellulosic fibres as well. They, analysed their colour yield and related results on colour fastness properties of natural dye extract from Shore Robusta bark. Wool was reported mordanted with metallic ions like Al (III), Cr (VI), Cu (II), Fe (II), Sn (II) and rare earths elements such as La (III), Sm (III) to colour extracted from beet sugar and the dyeing could satisfy colour fastness standards of BIS as reported by Mathur and Bhandari [136]. Agarwal et al. [137] optimised the mordant concentrations or mordant combinations and concluded that most beneficial shades that may be obtained with selective mordant combinations only such as 0.15% alum, 0.08% copper-sulphate and stannous-chloride mixture for optimised dyeing process with natural colour obtained from turmeric (*Curcuma longa*) for wool and also used 0.04% ferrous-sulphate and 0.06% potassium-dichromate for dyeing of mulberry silk fabric. But copper above a certain limit is not eco-friendly, and dichromate is not at all environmentally friendly. Hence, current research are focussed on the search of results of application of suitable bio-mordants or dual mordants applied as mixture or in sequences applying one after another for different natural dyeing. Hence, the study on effect of application of dual mordants preferably with bio-mordants combination in different ratio is needed to be explored for improving colour yield and colour fastnesses.

Irrespective of methods of mordanting, silk treated with potash-alum natural alum salt exhibited enhancement in colour, while silk mordanted with copper sulphate, potassium di-chromate and ferrous-sulphate evidenced very good colour fastness to light [138]. But, application of potassium-dichromate and copper-sulphate as mordant is not recommendable for environmental concern. Silk fabric mordanted with magnesium sulphate [139] produced lower depth of shade but at the same time, when it is mordanted with copper sulphate, it shows highest depth of shade. Copper above a certain limit is not environmentally safe. Das et al. [140] studied chemical alteration of cellulosic fabric in presence of acrylamide and $K_2S_2O_8$ to increase dyeability for natural dyes and to improve colour fastness to light and rubbing for these pre-treatment, although the said chemical modifications do not enhance washing fastness of such natural or synthetic dyed textile materials. Latter Das et al. [141] reported that application of aluminium sulphate and ferrous sulphate as mordants enhancing washing fastness of textile materials dyed with tea extract. Chan et al. [142] also reported that the effect of natural dye effluent on environment for wool dyeing with four varieties of tea and found that the protein fibres turned blackish in presence of mordant like ferrous sulphate. Gulrajani et al. [143, 144] reported the outcome of various mordanting agents on natural yellow colours such as kapila, onion, tesu, and dolu on different textiles.

Some mordants incorporated good wash stability to cotton, when it is dyed with golden rod using tin as a mordant and dyed with marigold by using chrome as a mordant and dyed with onion skins using alum and tin as mordants as reported by Vastrad et al. [145]. Bhattacharya and Lohiya [146] reported dyeing of cotton and polyester with catechu, nova red, turmeric and pomegranate rind as natural dyes with different eco-safe mordant.

Vat dyeing of cellulosic fibre in presence of iron (II) salt and complexing agent such as gluconic acid has been studied earlier by Chavan and Chakraborty [147], where they have reported to use of iron (II) salts complex with tartaric acid and citric acid as ligands at room temperature for its application on cotton using natural indigo. One report by Kumar and Bharti [148] showed that wash fastness and light fastness can be enhanced by using metal salts in presence of tannic acid or natural tannates for application on cotton fabrics. Pre-mordanted cotton yarns with potash-alum, ferrous-sulphate, potassium-dichromate and copper-sulphate are when treated with Acalypha extract [138] as natural dye, showed excellent colour fastness to light and wash, but potassium-dichromate and copper-sulphate are not eco-safe and ferrous sulphate always gives a darker grey to blackish shade.

A recent report described natural dyeing of cotton using gall nut extract as bio mordant (without any metallic mordants in combination) having natural tannates in extract for dyeing with extracts from babul bark as natural dye [149]. Dual pre mordanting with 15% concentration of harda and aluminium sulphate (in ratio 75:25) was applied in succession on cotton fabric and was subsequently dyed with aqueous extract from bark of babul [150] gave very good result in dye uptake/colour yield and colour fastness, proving that cotton fabric pre-treated with harda (myrobolan) as mordanting assistant (containing chebulinic acid) showed higher dye up take of respective natural dye for additional complexing of chebulinic acid of harda with metallic mordant like alum to form big giant complex of [Fibre-Harda-Metalic mordant-Natural dye] complex of much bigger size, improving wash fastness too. Double Pre-mordanting route also found to favourable for manjistha and other different natural dyeing of jute fabric as per earlier reports by Konar et al. [151].

Analysis of few tannin-based natural bio-mordants which can fix natural dyes on the fibres by forming a fibre-bio-mordant-natural dye complex/adduct, are already reported by Samanta [134], and also by Yusuf et al. [152]. Bio-mordants have advantages of its natural origin, eco safe and biodegradability criteria, free from toxicity causing no allergic reaction to human skin and sustainability [41]. Different bio-mordants (used with or without metallic mordant) have also been applied for eco-friendly natural dyeing of textiles in presence of their tannins/flavonoid content with mordantable capacity of poly-phenols/carboxylic acids which may assist dye fixation on fibres by forming larger (giant-sized) fibre-bio-mordant-metallic mordant (optional)-natural dye bigger giant complex. Examples of few such bio-mordants are gall-nut [149, 152]; myrobolan/harda [153], tamarind seed coat [154], pomegranate rind [155], aloe vera [156], lemon juice [157], mango bark [158] and banana sap [159], though use of different type of bio mordants and method of bio-mordanting with or without metallic mordants in combination having significant effect on rate and extent of colour yield and colour fastness to wash and photo fading behaviour due to presence of tannates.

According to one study, use of copper as metallic mordant with or without bio mordants showed high resistance to washing and photo fading; but stannous-chloride or alum combination with other bio mordants did not show such characteristics. However, light fastness/resistance to photo fading was found to improve after post-mordanting with copper ion, whereas, pre-mordanting with stannous-chloride or alum was found superior as reported by Gupta et al. [160].

Binary mordants combination with harda or tartaric acid was observed to be the very good mordant combination followed by tannic acid-harda and tartaric acid-tannic acid formulation as reported by Deo and Paul [161]. A synergistic effect of these mordant combination was found for use of any of the said binary combinations of mordants. According to this report, meta-mordanting exhibited superior results for harda-tartaric acid and tartaric acid-tannic acid formulation, while pre-mordanting showed good results for tartaric acid-harda formulation. Bains et al. [162] also studied the results of various combinations of mordants on colour fastness of wool dyed with golden drop is reported to show encouraging results.

Lots of literature are available on effect of different mordants or their combinations prior to natural dyeing on colour yield and fastness performance, particularly use of double or dual mordants with variation in their ratio, development of shade and colour yield and other colour interaction parameters [135, 163, 164], on cellulosic (cotton), ligno-cellulosic (jute), protein (silk and wool) and synthetic (nylon and polyester) fibres. But combination of Natural tannate based bio mordants along with natural metallic mordant i.e. potash alum or fitkari are not explored much, which need further study in this area for natural dyeing of cotton, jute, silk and wool textiles.

3.4 Methods of natural dyeing and effects of variation of dyeing process parameters

Ghorpade et al. [165] reported dyeing of cotton with sappan wood using ultrasound waves instead of simple aqueous bath dyeing. Tiwari et al. [166] studied natural dyeing of cotton with extract of tulsi leaves using ultrasonic waves. Lokhande et al. [167] reported dyeing of polyamide with three distinct natural dyes with different mordants by two different techniques (open bath and HTHP) and found HTHP dyeing is better than open bath dyeing. Bhattacharya and Lohiya [146] also utilised HTHP technique for dyeing cotton-polyester textiles with pomegranate rind, catechu and turmeric.

Tiwari et al. [166, 168] used the unconventional natural dyeing with alkanet root bark using both microwave and sonicator. Kamel et al. [169] studied colouring of wool fabric by the liquid extract of terminilla arjun fruit and chochineal mixture. Neetu and Jahan [170] reported the dyeing of combination of natural colours derived from onion skin and kilmora root. Samanta et al. [5] studied the use of mixture of turmeric and madder for cotton and observed synergistic effect for enhancing colour strength and improves washing fastness for mixed shades of turmeric and madder than that obtained for application of only turmeric on cotton. Compatibility of binary mixture of dyes for producing compound shades are studied by Samanta et al. [92, 97, 98, 164].

However, there are only sporadic report and very less scientific or technological reports on either statistical optimization/standardisation of dyeing process variables [149–151], studies on dyeing kinetics [85, 86], dye compatibility tests with conventional and newer colorimetric CDI based methods [92, 97, 98] and effect of bio mordants [149] and bio-finishes [134] for enhancement of colour strength, colour uniformity and colour fastness to wash by use of single and binary mixture of natural dyes on various natural textile fibres. Hence, there are ample scope of research in this area, particularly on standardisation of the dyeing process variables for natural dyeing using bio-mordants and also to study the effects of different bio-finishes for improving antimicrobial and UV-resistant, as wellness properties.

Computer aided colour matching system has become an integral part of textile dye houses as an important tool in textile processing unit for dyeing any textiles with particular class of synthetic dyes. The prediction of colour matching formulation using natural dyes is practically difficult and not that easy. Hence, it is not systemised yet for commercial applications for the following difficulties. And variation of colours for natural dyeing due to:

i. Varying shades, depending on source, extraction method of colour and non-reproducible nature.

ii. Lack of standardised dyeing process parameters and hence variation in colour after dyeing.

iii. Besides colour, the shade and hue is also much dependant on type and concentration of mordants.

iv. Usually for purified synthetic dyes, plot of dye concentration vs. *K*/*S* value is linear and *K*/*S* are additive, but for natural dyes, in most of the cases, this linearity is not maintained due many variations in the natural resourced materials and also due to additional process of mordanting etc. Hence separately linearization of data base, standardisation of mordanting and dyeing process conditions to obtain uniform and reproducible dyeing etc. are to be assured.

The above difficulties can be partly or fully eliminated by taking extracted purified natural dye source of known concentration only for calibration dyeing for preparation of correct database eliminating colour variations, using fixed type and concentration of mordants following optimised standard process of mordanting and dyeing, to obtain linear curves for purified natural dye concentrations vs. *K*/*S* values, applying linear least square curve fitting principles, and then one can use that data base for colour match prediction with known and purified natural dyes combination for colour match prediction with natural dyes allowing higher tolerances of colour differences.

3.5 Physico-chemical studies on dyeing process variables and dyeing kinetics

Different natural colours have different chemical constituents, which are determined by LC-MS or GC-MS analysis [149], while its extraction of colour components and finishing components or both depends on extraction media, temperature, pH, time, solvent and other conditions/factors of natural dye source like which part of the plant is taken, from where or which source the dye was extracted, area/soil of the field of cultivation/growth, atmospheric conditions of the growing areas, time of harvesting, extraction method or technology applied etc. Some researchers have reported discrete and piece-meal experimental inferences of laboratory testing regarding techniques and method of extraction including temperature, time, solvent-vegetal ratio, type of solvent used and extraction methods etc. Similarly colour strength produced with those extracted natural dyes depends on methods and types of mordanting, dyeing process variable parameters and nature of textile substrate etc. [171]. Hence, it is essential to study fibre-wise and dye-wise optimization of extraction parameters and standardisation/optimization of process variables of dyeing for different natural fibres to be dyed with different natural dyes using different single and double mordants and single and compatible binary mixture of dyes.

The effects of dye extraction mode with variation in pH, media, time and temperature used during extraction, and similarly for pre-mordanting stage, effect of type of mordant and mordant concentration and method of mordanting and mordanting conditions, and finally for Dyeing, effects of dye concentration and method of dyeing and variation in all dyeing process parameters are most important. For silk dyeing with some specific natural colourant effect of dyeing process parameters/conditions have been studied by Grover et al. [172], Dixit and Jahan [123] and Samanta et al. [149–151]. From the study made by Grover et al. [172], it was inferred that the acidic pH showed highest absorption for jatropa, lantana, hamelia and euphorbia dye, on the other hand kilmora and walnut exhibited better outcome in alkaline pH. The results shown in their various experiments lead to the optimization of a standard recipe for particular dye-mordant-fibre combination. Dumitrescu et al. [173] also studied the dyeing parameters which gave optimised and best result of colour yield and colour fastness on wool. The optimum concentration of colourant from beet sugar for wool dyeing was found to be 0.03 gm/gm of wool at pH 4, 5 at 97.5°C [39]. Deo et al. [174] described dyeing of cellulose and ligno-cellulosic materials with tea extract. Samanta et al. studied optimization of dyeing process variables [149–151] and compatibility [92, 97, 98] of binary mixtures of dyes applied on jute and cotton textiles with single and double mordants for different natural dyes. Das et al. [141] have reported that the colouring constituent of tea showed maximum attraction for both wool and silk at pH 2–4 in presence of ferrous-sulphate and aluminium-sulphate as mordants. Optimisation of dyeing process parameters for wool dyeing with natural dye derived from turmeric has been analysed by Agarwal et al. [175]. Gupta et al. [176] reported the kinetics and thermodynamics of fuglone dyeing and showed that linear isotherm for wool, human hair, silk, nylon and polyester indicated a partitioning dyeing mechanism. The slope of isotherms was raised with the rise of temperature for all cases. Both ΔH and ΔS value were found positive for all the dyeing. The apparent diffusion co-efficient was maximum for wool and minimum for silk. Samanta and Agarwal et al. [85, 86] have studied dyeing kinetics for dyeing jute and cotton with jackfruit-wood and red-sandal-wood natural colour.

A latest report on review of Scientific aspects and revival of incorporation of natural colours on textiles and a brief accounting of latest research done on natural dyes for textiles covering different scientific issues has been compiled and edited by Vankar [39] and also another two chapter contributed by Samanta [163, 164] in an edited book, covering all the aspects of natural dyeing including its extraction, characterisation and application on textiles.

A comprehensive current review on sources and composition of different natural dyes [135] and standardising method of application of natural colourants on different textiles and its characterisation with strategy for its revival are available in current literature [177, 178].

3.6 Colour fastness characteristics of natural dyes

Colour fastness is defined as the resistance of a material to alter its colour features or, transfer of its colour to adjacent materials in contact, or both, under different atmospheric condition and or any process like laundering, dry cleaning, etc. or exposing to various stimuli like heat, light etc. Fading means alteration/loss in colour depth after exposure to any environment/agency/process either by lightening or darkening of the shades.

Extensive works have been accomplished to increase the light fastness features of textiles dyed with natural colour. Cook [48] has reported a wide-range of review on various attempts taken for enhancing colour fastness of various natural dyes applied on several textile fibres by different techniques and methods. The said review also addressed after-treatments linked with tannin for enhancing the washing stability and colour fastness to light of such mordantable natural dyes suitable for cotton. Few of these chemical treatments might be relevant for specific natural colours and few are commonly applicable. Better alternative is to find out suitable natural resource based after-treatment, if possible, e g., to improve wash fastness of natural colourants by treatment of dyed fabrics with chitosan in acidic media and to improve light fastness, natural resource based after-treatment with antioxidant/UV resistant natural materilas like pomegranate rind extract or ashwagandha extract or Eucalyptus leaf or bark extract etc.

The light fastness is also associated with resisting the effect of UV light exposure initiated oxidation/reduction of natural/synthetic dyestuffs [49]. The ultraviolet light (UV) is a part of the electromagnetic radiation spectra having shorter wavelength (λ) than visible light. UV light may be divided into three groups e.g., UV-A (λ—320–400 nm), UV-B (λ—280–320 nm) and UV-C (λ—100–280 nm). As per law of physics, shorter wavelength (λ) has higher energy (E) level (as per Einstein's rule: $E = h\nu = [h \times c/\lambda)$ causing more detrimental effect on dyestuff or human skin. Luckily, high concentration of ozone in the stratosphere (around 15–30 km above the surface) absorbs UV-C radiation totally. UV-B is less than 1% of the radiation of the sun reaching to earth and is not much damaging, while UV-A (λ—320–400 nm) consists of around 5.6–6% of total radiation which reached to the earth surface. So, identified UV absorbing (UV-A component) natural resource based materials are known to have better UV protective action and hence after treatment with UV resistant dyes or finish materials can cause improving light fastness better than other dyes/finishes, not having this UV absorbing character.

Majority of the natural colours showed inferior to moderate wash stability and also inferior to moderate light fastness on UV light exposure (with respect to the comparable best synthetic dyes) and that is why, developed shades are sometimes lighter or different from their original colours. A comparative light fastness for a range of natural colourants was examined by Padfield and Landi [179], who also revaluated the development in qualitative fashion for selective natural dyes. These lightening or otherwise alteration in shade after washing for natural dyed wool textiles were measured quantitatively by Duff et al. [180], who evaluated the colorimetric analysis to note the changes of colour values in the Munsell scale and also with respect to the CIE colourimetric terms. Wool coloured with nine different natural colourants was exposed to Microscal MBTF fading lamp and finally the colour fastness ratings were found in line with proposition of Padfield and Landi [179] on exposure to daylight fading. During evaluation for light fastness (LF) as per blue wool standards, yellow natural colourants (old fustic and Persian berries) showed inferior light fastness with

1–2 rating; reds colours (cochineal with tin metallic mordant, alizarin with alum and metallic stannous mordant, lac with stannous metallic mordant) showed moderate LF with rating 3–4; indigo showed LF with rating 3–4 or 5–6 some times based on the mordant and logwood black used as natural colourants, showing LF of 4–5 with chromium containing mordant or 6–7 with other mordants. Gupta [181] reported that light fastness and all other colour fastness behaviour also depends on chemical constituent/structure of relevant natural dyes like ratanjot.

A mordant is required for majority of the natural colour components to be fixed on the textile substrate. Nature, type and concentrations of mordant has a significant role on wash stability and light fastness. The results of various mordants have a crucial role in fading of 18 different yellow natural dyes as studied by Crews (Crews P. C. 1982) [182]. Turmeric, marigold and fustic dyes have more tendency to fade than other natural dye of yellow shade. However, application of stannous metallic mordant and alum mordant combination showed higher fading than the mordants containing iron, copper or chromium. Hence, an inference may be drawn that the type of mordant is very crucial factor for determining the wash fastness (WF) and light fastness (LF) of textiles dyed with natural-colours. The LF of such natural colourants on wool has been compared with 'imported' dyes of similar shades by Duff et al. [180], again using Micro-scale MBTF fading lamp for simulated UV light exposure.

Oda reported [183, 184] the influences of various additives/add-ons on the photo-fading behaviour for carthamin natural dyes applied on cellulose acetate film and cotton cellulose yarns for improvement of light fastness. The presence of nickel hydroxy-aryl-sulphonates can suppress photofading intensity remarkably, while UV-absorbing dyes or finishes or such additives can retard the light fading rate remarkably. Cristea and Vilarem [185], Lee et al. [186] and Micheal et al. [187] have attempted to increase the LF of various natural fibre-based and natural dyed textile fabrics.

Duff et al. [180] studied wash fastness rating of few natural dyed old textiles with respective study of wash stability. Experiments were conducted under 50°C and at 20°C temperature with a washing combination suitable for preservation work for refurbishing of old textiles materials. Some colours showed remarkable alteration in their hue after washing with addition of small amounts of soda-ash or alkali in washing bath formulation, highlighting the importance of the pH for cleaning of any textiles containing natural colourants. In general, natural colourants (on wool) shows moderate wash stability, if evaluated as per ISO-II method.

Hofenk and Graaff [83] examined the nature of fading for natural dyes for both light and wash fastness targeting to retaining/restoration of the original colours of natural dyed-old church or museum textiles or painted flags for conservation. The effects of wash bath surfactant solution for conservator work for natural coloured textiles have been discussed by Hofenk and Graaff [83]. A liquor containing 1gpl of sodium-poly-phosphate was felt to be best as reported by Duff et al. [180] for test of wash stability of old natural dyed textiles remained in the old textiles of church.

Any small increment in cleaning efficiency attributed by the alkali, should be equalised to resist probable changes of shades of natural colours, apart from probable damage of protein fibres in alkaline pH conditions. Duff et al. [180] also examined the wash stability for natural dyes extracted from native Scottish locality/source as compared to imported natural dyes. The fastnesses of the logwood and indigo were found superior against other natural dyes like water-lily root and privet berries respectively when tested as per ISO-II method, but in comparison of imported vs. native natural colourants for yellow, reds, red/purples, greens and browns shades, the difference of wash fastness rating was observed to be minor in between these two groups.

Gulrajani et al. [188] described application of red-sandal wood extract to dye wool and nylon as compared to some other selective natural dyes determining all physico-chemical parameters and rate of dyeing and colour fastness properties. The said fabrics were coloured with four distinct natural colours (turmeric, myrobolan, madder, red sandalwood) with pre, post and simultaneous-mordanting techniques with Aluminium-sulphate mordant. Few samples were also coloured with a mixture of turmeric/madder or turmeric/red-sandalwood and myrobolan mixed with madder or red-sandalwood in various ratio to obtain a compound uncommon shade. Selected mordanted and natural coloured samples were post-treated with normal cationic dye fixing agents to improve wash fastness. Being a direct dye, Turmeric exhibited maximum colour strength during concurrent mordanting with single or combination of other colourants. Turmeric exhibited inferior wash stability, which was enhanced a little during application of a cationic dye fixing agent in after-treatment process or when applied in combination some other natural dyes of higher wash fastness with turmeric having lower rating of wash fastness, to obtain a compound shade of good washing fastness. Combined of turmeric with other natural dyes by simultaneous mordanting method exhibited good colour yield for development of darker shade and observed higher colour strength values than the calculated or expected values. Myrobolanin mixed with other dyes showed more prominent shades with higher colour strength value during its application by post-mordanting process. In concurrent mordanting process, myrobolan did not exhibit any synergistic effect with respect to observed and calculated K/S values, but post-mordanting gave the best results.

Konar et al. [189] studied application of dye extracted from tesu on jute fabric with varying aspects of mordants as well as with varying dyeing process variables to standardise this dyeing process. Application of selective pre-mordant (single and double mordants) on 6% H_2O_2 (50%) bleached jute fabric was carried out using myrobolan (harda) and eco-safe metallic salts mordants (like potash-alum or fitkari and aluminium-sulphate) and subsequent dyeing with aqueous extract of tesu (palash flower petal) under different dyeing conditions for optimization of the dyeing process variables. It was observed that 1st pre-mordanting with 20% myrobolan followed by 2nd mordanting with aluminium-sulphate (20%), in succession, is the most potential double pre-mordanting system rather than when used separately as single mordant, considering the results of textile related characteristics and colour yield for this dyeing. Effect of important process variables regarding dyeing (e.g., time, temperature, pH, MLR, concentration of mordant, dye and salt) on surface colour strength has been evaluated to determine the optimum dyeing conditions. Along with washing, rubbing and light fastness results. Generally, this dyeing is observed to be very sensitive to pH for selective fibre-mordant-dye combinations and it is observed here that dyeing at pH of 11 renders better colour yield and overall all round better colour fastness properties. Enhancement in washing and light fastness can be achieved with suitable post-treatment with cationic dye fixer and UV stabiliser compounds. Dyeing kinetics for dyeing tesu on jute was also reported by Konar et al. [86].

Samanta et al. [85, 190, 191], studied physico chemical parameters of natural dyeing on jute and cotton including its dyeing kinetics and also compatibility of a series of binary pairs of natural colourants by determining all the colour interacting parameters and colour fastness properties. Bleached jute and cotton fabrics (after consecutive premordanting with myrobalan and aluminium-sulphate) have been coloured with a definite ration and concentration of purified mixture of jackfruit wood (JFW) with other natural dyes, like manjistha (MJ), red sandalwood (RSW), marigold (MG), sappan wood (SW) and babul (BL) bark to get desirable combined shades with different colour strength (*K*/*S* values) and improved colour fastness. To increase the light fastness and wash-fastness of these binary mixture of natural dyed cotton fabrics, it need additional after-treatment with selected cationic dye-fixing agent along with an UV-absorbing compound, respectively. Post-treatment with 1% CTAB or 1% Sandofix HCF showed one level higher washing fastness for these dyed fabrics and 1% benztriazole treatment showed increment in LF rating approximate by one unit. Singhee and Samanta [192] has studied standardisation of dyeing process variables for application of Tesu as natural colour on silk fabric. Besides natural colouration, UV protective finishing action of anar peel (pomegranate ring) as a bio-dye cum bio-finishing material, applied on cotton khadi fabric, was reported by Sinnur and Samanta et al. [153] as a newer approach of bio-dyeing and bio-finishing, in which field, there are very few scanty reports on similar work with other bio-dyes and bio-finish combinations, need to explore further.

However, from consumer protection point of view. It was highly needed to standardise test methods for identification of exact natural dyes from such natural dyed textile substrate, to assure customers that the dyed product is natural dyed materials. With long painstaking effort for required R&D work and inter lab test trial report in this endeavour, a group of Indian researchers have developed and established dye-wise test methods for identification of specific natural dyes, e g. test method to differentiate between natural indigo and synthetic indigo for identifying natural indigo, which is published by BIS-India as new Indian standards for identification of natural Indigo [193]. A similar Standards on identification of natural madder differentiating it from synthetic alizarin colourant is also published by BIS as another new BIS standard [194] and later in a series, more than eight more natural dye test standards for identification of that specific individual natural dye extracting from that natural dyed textiles are published by BIS for the benefits of consumers by establishing some form of natural dye mark scheme, required both nationally and globally, for assurance of natural materials used.

4. Concluding remarks

Thus from the above said literature review, it is understood that there are still many lack/gaps in scientific studies on effects of different bio-mordants (such as tannin based natural complexes) and mordanting assistants (gallic acid from gall nut or chebulinic acid from harda-based natural complexes) and standardisation/optimization of dyeing process variables as well as effects of different natural resource/plant based after-treating or finishing compounds as dye fixatives/UV absorbers for improving wash fastness, Sun-light/UV light fastness and rubbing fastness besides attempts to increase antimicrobial and UV protection properties of such natural dyes by applying selected post treatments with different natural agents or eco-friendly synthetic agents. It is also necessary to understand the possible dye-fibre interactions and function of various pre and post treatments/additives for increasing colour yield (with respect to *K*/*S* values, uniformness of colour yield) by CV% of *K*/*S* values as well as antimicrobial grading and UV protection factor for various mordant–fibre-natural colourants application system particularly for cellulosic fibre (cotton) and protein fibres (like silk) when colouring with aqueous or aqua-alcoholic extract of chosen natural colourants. Hence, for revival of application of natural colourants on producing natural bio mordanting-bio dyeing and bio-finishing for producing natural resource based medical textiles and skin friendly high value apparel textiles.

However, one must remember that natural colours are not a replacement of man-made synthetic dyes. Natural dyes have their own customer base and any further enlargement of this customer base for natural-coloured products will not squeeze down the market of any man-made synthetic dyed textiles. So, further studies are required to meet the above-mentioned gaps for applying natural dyes and natural finishes on textile fabrics for improving wellness properties like antimicrobial, mosquito repellent, anti-odour and UV-resistant properties in apparel textiles.

Author details

Pubalina Samanta[1,2]*, Adwaita Konar[3] and Asis Muhopadhyay[4]

1 Department of Jute and Fibre Technology (DJFT), University of Calcutta, Kolkata, West Bengal, India

2 Department of Fashion and Apparel Design, Rani Birla Girls' College, Kolkata, WB, India

3 Government College of Textile Technology and Engineering, Serampore, West Bengal (Affiliated to MAKAUT, West Bengal), India

4 Department of Jute and Fibre Technology (DJFT), Institute of Jute Technology, University of Calcutta, Kolkata, West Bengal, India

*Address all correspondence to: samantapubalina@gmail.com; pubalina@gmail.com

References

[1] Gulrajani ML, Gupta D. Natural Dye and their Application to Textiles. Delhi: Dept of Textile Technology, Indian Institute of Technology; 1992. pp. 25-26, 50-55

[2] Samanta AK, Konar A. Technical Handbook on Natural Dye and Colouration. University of Calcutta: Dept. of Jute and Fibre Technology, IJT; 2012. pp. 97-113 (ch-4)

[3] Kamat SY, Alat DV. Natural dyes, a dyeing craft. Indian Textile Journal. 1990;**100**:66-69

[4] Samanta AK, Agarwal P. Application of natural dyes on textiles. Indian Journal of Fibre & Textile Research. 2009;**34**: 384-399

[5] Samanta AK, Singhee D, Sethia M. Application of single and mixture of selective natural dyes on cotton fabrics-a scientific approach. Colourage. 2003; **50**(10):29-42

[6] Chavan RB. Revival of natural dyes-A word of caution to environmentalists. Colourage. 1995;**42**(2):27

[7] Vankar PS. Handbook on Natural Dyes for Industrial Applications. Chapter 1. New Delhi, India: National Institute of Industrial Research; 2013. pp. 1-5

[8] Gulrajani ML. Present status of natural dyes. Indian Journal of Fibre & Textile Research. 2001;**1 & 2**(special issue):26

[9] Singh OP. Natural dyes: The pros and cons. Indian Textile Journal. 2000; **110**(4):42

[10] Samanta AK, Konar A. In: Kumbassar EPA, editor. Natural Dyes. London, UK: IntechOpen; 2011. pp. 29-56. ISBN-978-953-307-783-3

[11] Teli MD, Paul R. Dyeing of textiles with natural dyes. International Dyers. 2006;**191**(4):29-32

[12] SERI. Colour from Nature - Silk Dyeing Using Natural Dyes. New Delhi: SERI, Oxford and IBH Publishing Co. Pvt. Ltd.; 2002

[13] Mohanty BC, Chandranouli KV, Nayak ND. Natural Dyeing Processes of India. Ahmedabad, India: Calico Museum of Textiles; 1984. pp. 298-325

[14] Vankar PS, Shukla D, Wijayapala S, Samanta AK. Innovative silk dyeing using enzyme and *Rubia Cordifolia* extract at room temperature. Pigment and Resin Technology. 2017;**46**(4):96-302

[15] Hossain A, Saiful Islam AKM, Samanta AK. Pollution free dyeing on cotton fabric extracted from swieteniamacrophylla and Musa Acuminata as unpolluted dyes and Citrus limon (L) as unpolluted mordanting agent. Trends in Textile Engineering and Fashion Technology. 2018;**3**(2):1-8. DOI: 10.31031/TTEFT.2018.03.000558

[16] Hossain A, Samanta AK, Bhaumik NS, Vankar PS, Shukla D. Non-toxic coloration of cotton fabric using non-toxic colorant and non-toxic crosslinker. Journal of Textile Science & Engineering. 2018;**8**(5):1-8. DOI: 10.4172/2165-8064.1000374

[17] Hossain A, Samanta AK. Green dyeing on cotton fabric demodulated from *Diospyros malabarica* and *Camellia sinensis* with green mordanting agent. Latest Trends in Textile and Fashion Designing. 2018;**2**(2):1-8. DOI: LTTFD. MS.ID 000132

[18] Hossain A, Samanta AK, Bhaumik NS, Vankar PS, Shukla D.

Organic Colouration and antimicrobial finishing of organic cotton fabric by exploiting distillated organic extraction of organic *Tectona grandis* and *Azardiracchta indica* with organic mordanting compare to conventional inorganic mordants. International Journal of Textile Science and Engineering. 2018;**1**(1):1-12. DOI: 10.29011/IJTSE-113/100013

[19] Patra SK, Nanda B, Nayak A, Tiwari NB. Application of natural dyes (review article). Colourage. 2000;**47**(8): 17-22

[20] Grierson S, Duff DG, Sinclair RS. Natural dyes of the Scottish highlands. Textile History. 1985;**16**(1):23-43

[21] Samanta AK, Konar A, Chakrabarti (Mukherjee) S, Datta S. Effect of different mordants, extraction conditions and dyeing process variables on colour interaction parameters and colour fastness properties in dyeing of jute fabric with manjistha, a natural dye. Journal of the Institution of Engineers (India)-Textile & Chemical Engineering. Series-E. 2010;**91**:7-15

[22] Samanta AK, Datta M, Datta S. Dyeing of jute fabric with binary mixtures of catechu and other natural dyes: Study on colour performance and dye compatibility. Journal of Materials Sciences and Applications, AASCIT-USA. 2015;**1**(5):221-238

[23] Dedhia EM. Natural dyes. Colourage. 1998;**16**(3):45-53

[24] Gupta D, Jain A, Panwar S. Anti-UV and antimicrobial properties of some natural dyes on cotton. Indian Journal of Fibre & Textile Research. 2005;**30**(6): 190-195

[25] Gupta G. Mechanism of dyeing synthetic fibres with natural dyes. In: Gupta D, Gulrajani ML, editors. Proceedings of 1st Convention on Natural Dyes; 9th to 11th December, 1999, Dept. of Text Tech., IIT Delhi. Delhi, India: Replica Press Pvt Ltd; 1999. pp. 121-124

[26] Samanta AK, Konar A, Datta M, Datta S. Natural dyeing of jute fabric with catechu. International Journal of Applied Engineering Research. 2014; **9**(4):451-467

[27] Dixit S, Jahan S. Colourfastness properties of euphorbia (*Euphorbia continifolia*) leaves dye on silk fabric. Man-Made Textiles in India. 2005;**48**(7): 252-258

[28] Vankar PS, Tiwari V, Ghorpade B. Supercritical fluid extraction of natural dye from eucalyptus bark used for cotton dyeing in microwave and sonicator. In: Gupta D, Gulrajani ML, editors. Proceedings of 2nd Convention on Natural Dyes; 17th-18th December, 2001, Dept. of Textiles Tech., IIT Delhi. Delhi, India: Replica Press Pvt Ltd; 2001. pp. 53-55

[29] Deo HT, Paul R. Ultrasonic dyeing of cationized cotton fabric with natural dye using potassium alum in combination with Harda and tartaric acid. Indian Journal of Fibre & Textile Research. 2000;**25**(3):17-23

[30] Ghorpade B, Darrekar M, Vankar PS. Eco-friendly cotton dyeing with sappan-wood dye using ultrasound energy. Colourage. 2000;**47**(1):27-32

[31] Vankar PS, Shanker R. Ecofriendly ultrasonic natural dyeing of cotton fabric with enzyme pretreatment. Desalination. 2008;**230**(1/3):62-69

[32] Tiwari V, Vankar PS. Unconventional natural dyeing using microwave and sonicator with alkanet

root bark. Asian Textile Journal. 2001;**10** (5/6):54-60

[33] Lokhande HT, Dorugade VA. Dyeing nylon with natural dyes. American Dyestuff Reporter. 1999;**88**(2):29-33

[34] Bhattacharya N. Natural dye – Its authenticity and identification. In: Gupta D, Gulrajani ML, editors. Proceedings of 1st Convention on Natural Dyes; 9th to 11th December, 1999, Dept. of Text Tech., IIT Delhi. Delhi, India: Replica Press Pvt Ltd; 1999. p. 134

[35] Mathur PJ, Mehta A, Karnawar R, Bhandari CS. Use of neem bark as wool colourant – Optimum conditions of wool dyeing. Indian Journal of Fibre & Textile Research. 2003;**28**(1):94

[36] Dev VG, Venugopal J, Sudha S, Deepika G, Ramkrishna S. Dyeing and antimicrobial characteristics of chitosan treated wool fabrics with henna dyes. Carbohydrate Polymers. 2009;**75**(4): 646-650

[37] Patel KJ, Patel BH, Naik JA, Bhavsar AM. Eco-friendly dyeing with tulsi leaf extract. Man-Made Text India. 2002;**45**(11):420-424

[38] Yusuf M, Shabbir M, Mohammad F, Khan MA. Eco-dyeing of wool with *Rubia cordifolia* root extract: Assessment of the effect of acacia as biomordant on colour and fastness properties. Textiles and Clothing Sustainability. 2016;**2** (10):1-9

[39] Vankar PS. Natural Dyes for Textiles, The Textile Institute Book Series. UK: Elsevier; 2017. pp. 17-26, 89-103, 141-166

[40] Vankar PS, Shanker R. Ecofriendly pretreatment of silk fabric for dyeing with delonixregia extract. Coloration Technology. 2009;**125**(3):155-160

[41] Vankar PS, Shanker R, Mahanta D, Tiwari SC. Ecofriendly Sonicator dyeing of cotton with rubia cordiafolia Linn using biomordants. Dyes and Pigments. 2008;**76**(1):207-212

[42] Pant S, Mangal R. Modifying the dyeability of nylon for natural dye lac. Manmade Text India. 2011;**39**(2):57-62

[43] Sankar R-M, Bhowmik L. Studies on application of some vegetable dyes on cellulosic and ligno-cellulosic fibre. Manmade Text India. 2006;**49**(4): 142-146

[44] Singh S, Jahan S, Gupta KC. Optimisation for procedure for dyeing silk with natural dye - madder roots. Colourage. 1993;**40**(8):33-36

[45] Zhang RP, Cai ZS, Zhu H, Yu BL, Chen HJ. Influences of enzyme on the dyeing performance of wool with natural pigment sappanwood. Wool Textile Journal. 2011;**12**:2-7

[46] Gahlot M, Kaur S. Rebirth of natural dyes. Indian Textile Journal. 1996; **106**(5):46-48

[47] Vastard J, Shailaja D, Mamatha A. Dye's colour fastness. Indian Textile Journal. 1999;**109**(7):68-73

[48] Cook CC. Aftertreatments for improving the fastness of dyes on textile fibre. Review of Progress in Coloration. 1982;**12**(1):78-89

[49] Yadav R, Battacharya N. Effect of acacia catechu on UV protection of cotton, polyester and P/C blend fabrics. Colourage. 2005;**52**(6):49-54

[50] Cristea D, Vilarem G. Improving light fastness of natural dyes on cotton

yarn. Dyes & Pigments. 2006;**70**: 238-244

[51] Miguel MG, Neves MA, Antunes MD. Pomegranate (*Punica granatum* L.): A medicinal plant with myriad biological properties—A short review. Journal of Medicinal Plants Research. 2010;**4**(25):2836-2847

[52] Adeel S, Ali S, Bhatti IA, Zsila F. Dyeing of cotton fabric using pomegranate (*Punica granatum*) aqueous extract. Asian Journal of Chemistry. 2009;**21**(5):3493-3499

[53] Weerakkody P, Jobling J, Infnte MMV, Rogers G. The effect of maturity, sunburn and the application of sunscreens on the internal and external qualities of pomegranate fruit grown in Australia. Scientia Horticulturae. 2010; **124**(1):57-61

[54] Tayel AA, El-Tras WF. Anticandidal activity of pomegranate peel extract aerosol as an applicable sanitizing method. Mycoses. 2009;**53**:117-122

[55] Sentthil Kumar CS, Dhinakaran M. Extraction and application of natural dyes from orange peel and lemon peel on cotton fabrics. International Research Journal of Engineering and Technology. 2017;**4**(5):237-238

[56] Garifield S, Mauve WW. Natural Dyes for Textiles. New York: Norton and Company; 2001. pp. 55-65

[57] Sinnur HD, Samanta AK, Verma DK, Kaware R. Standardization of mordant and dyeing process variables for dyeing of cotton khadi fabric with Indian madder as natural dyes. Indian Journal of Natural Fibres. 2017;**4**(1):21-38

[58] Brian GJ. Dyeing what comes naturally to the dye house. Society of Dyers and Colourists. 1998; **114**(4):240

[59] Badri BM, Burkinshaw SM. Dyeing of wool and nylon 6.6 with henna and lawsone. Dye & Pigments. 1993; **22**(1):15

[60] Alam MM, Rahman ML, Haque MZ. Extraction of henna leaf dye and its dyeing effects on textile. Bangladesh Journal of Scientific and Industrial Research. 2007;**42**(2):217

[61] Shah NC. Traditional use of turmeric (*Curcuma longa*) in India. Journal of Medicinal and Aromatic Plant Sciences. 1997;**19**:948

[62] Indi YM, Chinta SK. Application and properties of natural dye on cotton - phyllanthusreticulatus. Colourage. 2008; **55**(6):52

[63] Yoshizumi K, Crews PC. Characteristics of fading of wool cloth dyed with selected natural dyestuffs on the basis of solar radiant energy. Dyes and Pigments. 2003;**58**:197

[64] Vankar PS, Shanker R, Dixit S, Mahanta D, Tiwari SC. Characterisation of the colorants from leaves of *Bischofia javanica*. International Dyer. 2008; **192**(3):11

[65] Saha P, Datta S. Dyeing of textile fibre using Marigold flower as floral dye. Colourage. 2008;**55**(5):58

[66] Gahlot M, Fatima N, Papna N. Jatropha flowers; a natural colourant for dyeing of cotton, wool and silk. Colourage. 2008;**52**(2):96-200

[67] Sidhu SP, Grewal JK. Silk dyed with goldendrop (Onosmaechiodes) dye. Colourage. 2008;**55**(3):51

[68] Mahale G, Sakashi, Sunanda RK. Arecanut - A natural colourant for silk. Colourage. 2008;**46**(4):54

[69] Vankar PS, R. Shanker 'ultrasonic dyeing of cotton and silk with *Nerium oleander* flower'. Colourage. 2008; **55**(4):51

[70] Purohit A, Das R, Nayak A, Nanda B, Das NB. Natural dyes from the waste leaves of artocarpusheterophyllus. Colourage. 2009;**56**(5):89

[71] Teli MD, Shah R, Sabale AG. Dyeing of nylon with natural colourants. Colourage. 2010;**57**(3):96

[72] Chairat M, Rattanaphani V, Bremner JB, Rattanaphani S, Parkins DF. An absorption investigation of the interaction of lac dye with metal ions. Dyes and Pigments. 2004;**63**(2):15

[73] Javali UC, Sreenivasa, Radhalakshmi YC. Application of natural dye-annatto on mulberry silk. Colourage. 2009;**50**(2):32

[74] Sakata K, Katayama A. Dyeing of silk fabrics with the pigments separated from a powdered dry Indian madder. Journal of Sericultural Science of Japan. 1996; **63**(3):170

[75] Sudhakar R, Gowda KNN. Dyeing of silk with flower extract of *Spathodea campanulata*. Man-Made Textiles in India. 2005;**48**(7):255

[76] Rattanaphani SS, Chairat M, Rattanaphani JB. An adsorption and kinetic study of lac dyeing on silk. Dyes and Pigments. 2007;**72**(1):88

[77] Patel BH, Bhatia KB, Parekh UD. Environmental-friendly & cost-effective method to create various shades on silk. Indian Silk. 2005;**44**(6):24

[78] Perkin AG, Everest AE. The Natural Organic Colouring Matter. London: Longmans Green; 1918. pp. 35-64

[79] Sahid M, Mohammad F. Recent advancements in natural dye applications: A review. Journal of Cleaner Production. 2013;**53**:310-331

[80] Mayer F, Cook AH. Chemistry of Natural Colouring Matters'. 3rd ed. Reinhold; 1943. pp. 45-65

[81] Thomson RH. Naturally-occurring Quinones. 2nd ed. Academic Press Inc., Springer; 1987. pp. 15-22

[82] Farris RE. Natural dyes. In: Kirk-Othmer Encyclopedia of Chemical Technology. 3rd ed. Vol. 8. New York: John Wiley and Sons; 1991. pp. 351-373

[83] Hofenk de Graaff JH. Conservation-restoration of church textiles and painted flags. In: Ágnes T-B, editor. Proceedings of 4th Int Restorer Seminar, 2-10 July 1983, Veszprém, Hungary. Vol. 1 and 2. Budapest: National Centre of Museums; 1983-1984. pp. 219-228

[84] Samanta AK, Agarwal P, Konar A, Datta S. Characterisation and application of purified selective natural dyes on jute, International Dyer. 2008;**192**(4): 25-32

[85] Samanta AK, Agarwal P, Datta S. Physico-chemical studies on dyeing of jute and cotton fabrics using jackfruit wood extract as natural dyes: Part-II: Kinetic and thermodynamic studies. Indian Journal of Fibre & Textile Research. 2008;**33**:66-72

[86] Konar A, Samanta AK, Datta S. Dyeing of jute fabric with tesu extract: Part-II: Thermodynamic parameters and kinetics of dyeing. Indian Journal of Fibre & Textile Research. 2012;**37**: 172-177

[87] McGovern PE, Lazar J, Michel RH. The analysis of indigoid dyes by mass spectrometry. Journal of the Society of Dyers and Colourists. 1990;**106**(1):22

[88] Wong ON. The identification of natural dyes in pre-Columbian Andean textiles by mass spectrometry [MA thesis]. Cambridge, MA, USA: Massachusetts Institute of Technology; 1977. p. 20

[89] Korenzvi C. HPLC analysis of the natural scale insect, madder and indigoid dyes. Journal of the Society of Dyers and Colourists. 1994;**110**(9):273

[90] Guinot P, Roge A. Dyeing plants screening approach to combine past heritage and present development. Colouration Technology. 2006;**122**(2): 93-101

[91] Samanta AK, Agarwal P, Datta S. Physico-chemical studies on dyeing of jute textiles with natural dyes extracted from red sandal wood. Journal of Institution of Engineers (India)—Textile Engineering. 2006;**87**:16-26

[92] Samanta AK, Agarwal P. Application of mixture of red sandal wood and other natural dyes for dyeing of jute fabric-studies on dye compatibility. International Dyer. 2008;**192**(3):37-41

[93] Mathur JP, Metha A, Kanawar R, Bhandari CS. Use of neem bark as wool colourant – Optimum conditions of wool dyeing. Indian Journal of Fibre & Textile Research. 2003;**28**(1):94

[94] Gulrajani ML, Gupta D, Roy-Maulik S. Studies on dyeing with natural dyes – Part III – Dyeing of ratanjot dye on nylon and polyester. Indian Journal of Fibre & Textile Research. 1999;**24**(4): 294

[95] Shankar R, Vankar PS. Dyeing wool with gomphrenaglobosa flowers. Colourage. 2005;**52**(4):35

[96] Bhuyan R, Saikai CN, Das KK. Extraction and identification of colour components from barks of *Musops elengi* and *Terminilia arjuna* and evaluation of their dyeing characteristics on wool. Indian Journal of Fibre & Textile Research. 2004;**29**(12):470

[97] Samanta AK, Agarwal P, Datta S. Dyeing of jute with binary mixtures of jackfruit wood and other natural dyes: Study on colour performance and dye compatibility. Indian Journal of Fibre & Textile Research. 2008;**33**:171-180

[98] Samanta AK, Priti Agarwal, Deepali Singhee Siddartha Datta, Application of single and mixtures of red sandal wood and other natural dyes for dyeing of jute fabric: Studies on colour parameters/ colour fastness and compatibility. Journal of the Textile Institute. 2009;**100**:565-587

[99] Blanc R, Espejo T, Montes AL, Toress D, Crovetto G, Navalon A, et al. Sampling and identification of natural dyes in historical maps and drawings by liquid chromatography with diode-array detection. Journal of Chromatography. 2006;**1122**(1–2):105

[100] Zin N W M S Moe, Purification of natural dyestaff extracted from mango bark for application in protein fibres - Proceedings of World Academy of Science, Engineering and Technology, 36, 2008, p. 540-544,

[101] Walker C, Needles HL. Historic Teles and paper materials. Advances in Chemical Series, Engineering. 1986;**175**:212

[102] Son YA, Hong JP, Kim TK. An approach to the dyeing of polyester fibre using indigo and its extended wash fastness. Dyes and Pigments. 2004; **61**(3):263-272

[103] Hemlatha J, Vashantha M. Ecofriendly dyeing with natural dye - Arcea nut; enhancing colour fastness

with natural mordants(Myrobolan, Lodhra and Pomogrenate) and increasing the antibacterial activity. Archieves of Appliaed Science Research. 2016;**8**(8):1-7

[104] Mondhe PS, Rao JT. Studies on jatrophacurcas seed oil incorporated alkyl dyes part I. Colourage. 1993;**43**(5):43

[105] Zippel E. Analysis of dye colourants and coloured textiles. Review of Progress in Coloration. 2004;**34**(1):1-11

[106] Joshi M, Purwar R. Developments in new processes for colour removal from effluent. Review of Progress in Coloration. 2004;**34**(2):58

[107] Khan MA, Khan M, Srivastav PK, Mohammad F. Extraction of natural dyes from cutch, ratanjot and madder. Colourage. 2006;**56**(1):61

[108] Mishra, Patni V. Extraction and application of dye extracted from eriphyid leaf galls of QuercuLeucotrichophora _A Himalayan Blue jack oak. African Journal of Biochemistry Research. 2011;**5**(3): 90-94

[109] Jain H. Dyeing of cotton fabrics with extract of jamun tree and its bye products using natural mordants. Colourage. 2013;**20**:40-43

[110] Benencia F, Courreges MC. Anti-viral activity of sandalwood oil against viruses 1 and 2. Phytomed. 1999;**6**(2):119

[111] Singh M, Bhattacharya N, Gupte VC. Determination of compatibility of reactive dyes in mixtures by chromatic diagram. Colourage. 2006;**62**(11):71

[112] Ferreria ESB, Hulme AN, Macnab H, Ouve A. The constituents of historical textile dyes. Chemical Society Review. 2004;**33**:329-336

[113] Haulak JP, Churrier B, Nargues M. HPLC analysis of elagic acid in European oak wood (*Quercus robur* L). HolzForschum. 1992;**46**:87-89

[114] Shenai VA. Vegetable dyeing research in India. Colourage. 2002; **49**(10):29

[115] Dayal R, Dobhal PC. Natural dye from Shorea robusta Bark. In: Gupta D, Gulrajani ML, editors. Proceedings of 2nd Convention on Natural Dyes; 17th-18th December, 2001, Dept. of Textiles Tech., IIT Delhi. Delhi, India: Replica Press Pvt Ltd; 2001. p. 47

[116] Ray MS, Pradhan SC. Dyeing of wool with Hinjal bark, jujube bark and Himalayan rhubarb. Man-Made Textiles in India. 2005;**48**(10):396

[117] Pan NC, Chattopadhyay SN, Day A. Dyeing of jute with natural dyes. Indian Journal of Fibre & Textile Research. 2003;**28**(3):339-342

[118] Verma N, Gupta NP. Wattle bark – A source of dyes for woollens. Colourage. 1995;**42**(7):27

[119] Saxena S, Iyer V, Shaikh AI, Shenai VA. Antioxidant activity of leaves and fruits of cultivated conifers in Iran. Colourage. 1997;**11**(3):23

[120] Sarkar D, Mazumdar K, Datta S, Sinha DK. Application of natural dyes from marigold flowers on cotton, wool and silk. Journal of the Textile Association. 2005;**66**(2):67

[121] Sarkar D, Mazumdar K, Datta S. Solvent extraction of yellow colouring matter from natural source; flower and its application on hydrophobic substances. Man-Made Textiles in India. 2006;**49**(1):19

[122] Deo HT, Paul R. Dyeing of ecru denim with onion extract using natural mordant combination. Indian Journal of Fibre & Textile Research. 2000;**25**(6):152

[123] Dixit S, Jahan S. Colour fastness properties of Euphorbia (*Euphorbia continifolia*). Man-Made Textiles in India. 2005;**48**(7):252

[124] Sudhakar R, Ninge KN, Padaki NV. Natural dyeing of silk with nut extract of areca catechu. Colourage. 2006; **53**(7):61

[125] Radhika D, Jacob M. Dye from jatropa seed. Indian Textile Journal. 1999;**109**(7):30

[126] Hagerman AE, Wilson TC. Quantitative determination of ellagic acid. Journal of Agricultural and Food Chemistry. 1990;**38**:1678-1683

[127] Nanda B, Nayak A, Das NB, Patra SK. Utilization of natural dyes from plant waste. In: Gupta D, Gulrajani ML, editors. Proceedings of 2nd Convention on Natural Dyes; 17th-18th December, 2001, Dept. of Textiles Tech., IIT Delhi. Delhi, India: Replica Press Pvt Ltd; 2001. p. 85

[128] Bhattacharya SK, Chatterjee SM, Dutta C. Effects of mordants on polymide fabrics dyeing with some ecofriendly natural colourants. Man-Made Textiles in India. 2004;**45**(8):85

[129] Onal A. Extraction of dyestaff from onion (*Allium cepa* L) and its dyeing properties in the dyeing of wool, feathered leather and cotton. Turkey Journal of Chemistry. 1996;**20**:194-204

[130] Agarwal A, Paul S, Gupta KC. Effect of mordant on natural dyes. Indian Textile Journal. 1993;**103**(4):110

[131] Houlton N, Raja, Kala J. Dyeing of jute cotton blends with grape skin waste. Journal of Textile Association. 2005; **66**(3):117

[132] Teli D, Adivarekar RV, Pardeshi PD. Dyeing of pretreated cotton substrate with tea extract. Colportage. 2002;**49**(10):23-26

[133] Padmaja A, Jacob M. Application of hibiscus flower dye extract on silk. The Textile Industry and Trade Journal. 1998; **36**(7):94-98

[134] Samanta AK. Bio-dyes, bio mordants and bio-finishes: Scientific analysis for its application on textiles. In: Samanta AK, Awwad N, Algarni HM, editors. Chemistry and Technology of Natural and Synthetic Dyes and Pigments. London, UK, Croatia & London UK: IntechOpen; 2020. pp. 3-37

[135] Samanta AK, Deepali S. Sources, application and analysis of natural colorants: An Indian perspective. In: Bechtod T, Manian AP, Pham T, editors. Handbook of Natural Colorants. 2nd ed. UK: Wiley; 2023. pp. 103-143. (ch 8)

[136] Mathur JP, Bhandari CS. Use of beet sugar as wool colourant. Indian Journal of Fibre & Textile Research. 2001;**26**(1):313

[137] Agarwal A, Goel A, Gupta KC. Optimisation of dyeing process for wool with natural dye obtained from turmeric (*Curcuma longa*). Text Dyers Print. 1992; **25**(10):28

[138] Mahale G, Sakshi G, Sunanda RK. Silk dyed with Acalypha (*Acalypha wilkesiana*) and its fastness. Indian Journal of Fibre & Textile Research. 2003;**28**(3):86

[139] Ray Maulik S, Pal P. Effect of mordants on colour uptake and fastness

properties of selected natural dyes on silk. Man-Made Textiles in India. 2005; **48**(1):19

[140] Das D, Bhattacharya SC, Sankar RM. Modification of cotton fabric with acrylamide in presence of $K_2S_2O_8$ for improving dyeability of natural dyes. Indian Journal of Fibre & Textile Research. 2006;**31**(12):559

[141] Das D, Bhattacharya SC, Sankar RM. Dyeing of wool silk with tea. International Journal of Tea Science. 2005;**4**(3&4):17

[142] Chan PM, Yuen CWM, Yeung KW. The effect of natural dye effluent on environment. Textile Asia. 2000;**31**(2):28

[143] Gulrajani ML, Gupta D, Agarwal V, Jain M. Some studies on natural yellow dyes—Part 2. Indian Textile Journal. 1992;**102**(5):78

[144] Gulrajani ML, Gupta D, Agarwal V, Jain M. Some studies on natural yellow dyes—Part 3. Indian Textile Journal. 1992;**102**(6):76

[145] Vastard J, Shailaja D, Mamatha A. Dye's colour fastness. Indian Textile Journal. 1999;**109**(7):68

[146] Bhattacharya N, Lohiya N. Dyeing of cotton and polyester fiber with pomegranate rind, catechu, nova red and turmeric. Asian Textile Journal. 2002; **11**(1):70

[147] Chavan RB, Chakraborty JN. Dyeing of cotton with indigo using iron (II) salt complexes. Colouration Technology. 2001;**117**(2):88

[148] Kumar V, Bharti BV. Eucalyptus yields dye. Indian Text Journal. 1998; **109**(5):18

[149] Dhanania Y, Singhee D, Samnata AK. Optimization of dyeing process variables for eco-friendly dyeing of cotton fabric with babul bark extract as a natural dye and gallnut extract as a bio-mordant. Journal of Natural Fibres. 2022;**19**(13):5478-5495

[150] Sinnur HD, Samanta AK, Verma DK. Standardization of dyeing process variables for dyeing of cotton khadi fabric with aqueous extract of babul bark (*Acacia nilotica* L.). Journal of Institution of Engineers (India)–Series-E (Chemical and Textile Engineering). 2018;**99**(2):187-207

[151] Samanta AK, Konar A, Chakrabarti (Mukherjee) S, Datta S. Effect of different mordants, extraction conditions and dyeing process variables on colour interaction parameters and colour fastness properties in dyeing of jute fabric with Manjistha, a natural dye. Journal of Institution of Engineers (India)-Textile Engineering. 2010;**91**:7-15

[152] Yusuf M, Shabbir M, Mohammad F, Khan MA. Eco-dyeing of wool with rubia cordifolia root extract: Assessment of the effect of acacia as biomordant on colour and fastness properties. Textiles and Clothing Sustainability. 2016;**2**(10):1-9

[153] Sinnur HD, Samanta AK, Verma DK, Runali K. Studies on colouration and UV protective action of anar peel (pomegranate ring) as an effective natural colourant for cotton khadi fabric. Journal of The Institution of Engineers (India): Series E. 2017, 2017; **99**(1):9-26

[154] Prabhu KH, Teli MD. Eco-dyeing using *Tamarindus indica* L. seed coat tannin as a natural mordant for textiles with antibacterial activity. Journal of Saudi Chemical Society. 2014;**18**(6): 864-872

[155] Ismal OE, Yildirim L, Ozdogan E. Use of almond shell extracts plus

biomordants as effective textile. Journal of Cleaner Production. 2014;**70**(9):61-67

[156] Zubairu A, Mshelia YM. Effects of selected mordants on the application of natural dye from onion skin (*Allium cepa*). Science & Technology. 2015;**5**(2): 26-32

[157] Janani L, Winifred D. Suitability of dyes from mulberry and coffee leaves on silk fabrics using eco-friendly mordants. International Journal of Scientific and Research Publications. 2013;**3**(11):1-4

[158] Wangatia LM, Tadesse K, Moyo S. Mango bark mordant for dyeing cotton with natural dye: Fully eco-friendly natural dyeing. International Journal of Textile Science. 2015;**4**(2):36-41

[159] Barhampurkar S, Bhat P, Kumar A, Purwar R. Studies of banana Sap used as mordant for natural dyes. International Journal on Textile Engineering and Processes. 2015;**1**(4):55-59. ISSN 2395 3578

[160] Gupta D, Gulrajani ML, Kumari S. Light fastness of naturally occurring anthraquinone dyes on nylon. Colouration Technology. 2004;**120**:205

[161] Deo HT, Paul R. Dyeing of ecru denim with onion extract using natural mordant combination. Indian Journal of Fibre & Textile Research. 2000;**25**(6):152

[162] Bains S, Kaur K, Kang S. Colour fastness properties of dyeing wool with Goldendrop (Onosmaechioides) dye using combination of mordants. Colourage. 2005;**52**(5):51

[163] Samanta AK. Application of natural dyes to cotton and jute textiles: Science and technology and environmental issues. In: Yusuf M, editor. Handbook of Renewable Materials for Coloration and Finishing. Chapter 11. Hoboken, NJ, USA: John Wiley & Sons Inc; 2018. pp. 213-278

[164] Samanta AK, Konar A. Dyeing of textiles with natural dyes. In: Perrin E, Kumbassar A, editors. Natural Dyes. London, UK, London & Croatia, UK: InTech International Publisher; 2011. pp. 29-56. ISBN-978-953-307-783-3

[165] Ghorpade B, Tiwari V, Vankar PS. Eco-friendly cotton dyeing with sappan-wood dye using ultrasound energy. Asian Textile Journal. 2000;**9**(3):68

[166] Tiwari V, Ghorpade B, Mishra A, Vankar PS. Ultrasonic dyeing of cotton fabric with aqueous extract of ecliptaalba. New Cloth Market. 2000; **14**(1):23

[167] Lokhande HT, Dorugade VA, Naik SR. Colour strength and colour fastness of flax fabrics dyed with natural colourants. American Dyestuff Reporter. 1998;**87**(6):40

[168] Tiwari V, Ghorpade B, Vankar PS. Ultrasonic dyeing with impatientsbalsamina using ecofriendly mordants on cotton. Colourage. 2000; **47**(3):21

[169] Kamel MM, El Zawahry MM, El-Ghafar FA. Dyeing wool fabrics with a mixture of natural dyes. Journal of Textile Association. 2001;**62**(4):167

[170] Neetu S, Jahan S. Dyeing wool by a combination of natural dyes obtained from onion skin and kilmora roots. Colourage. 2003;**50**(1):43-44

[171] Pant S, Mangal R. Modifying the dyeability of nylon for natural dye lac. Man-Made Text India. 2011;**39**(2):57

[172] Grover E, Sharma A, Rawat B. Dyeing of silk with natural dyes. International dyers. 2005;**190**(10):9

[173] Dumitrescu I, Mocioiu AI, Mocioiu AM. Natural compounds for

heathy clothes from vegetable wastes. World Textile Abstract. 2005;**56**(4):9

[174] Deo HT, Desai BK. Dyeing of cotton and jute with tea as natural dye. Colourage. 2006;**48**:33-38

[175] Agarwal A, Garg A, Gupta K. Development of suitable dyeing process for dyeing of wool with natural dye henna (lawsoniainerma). Colourage. 1992;**39**(10):43

[176] Gupta D, Gulrajani ML, Kumari S. Light fastness of naturally occurring anthraquinone dyes on nylon. Colouration Technology. 2004;**120**(3):205

[177] Pubalina S. A review on characterization and standardization for application of natural dyes on natural textiles: Part−1. Indian Journal of Natural Fibres (IJNF). 2019;5(2):53-63

[178] Pubalina S. A review on application of natural dyes on textile fabrics and its revival strategy. In: Samanta AK, Awwad NS, Algarni HM, editors. Book on Chemistry and Technology of Natural and Synthetic Dyes and Pigments. London, UK, Croatia and London UK: IntechOpen; 2020. pp. 75-98. (Ch-3)

[179] Padfield P, Landi S. The light fastness of the natural dyes. Studies in Conservation. 1966;**11**:161-196

[180] Duff DG, Sinclair RS, Stiriling D. The fastness to washing of some natural dyestuffs on wool. Studies in Conservation. 1977;**22**:161-116

[181] Gupta D. Dyeing of Ratanjot dye on nylon & polyester. Colourage. 1999; **46**(7):35

[182] Crews PC. The influence of mordant on the lightfastness of yellow natural dyes. Journal of the American. Institute for Conservation. 1982;**21**: 43-58

[183] Oda H. Improvement of light fastness of natural dyes. Colouration Technology. 2001;**117**(4):204

[184] Oda H. Improving light fastness of natural dyes on cotton yarn. Colouration Technology. 2001;**117**(5):254

[185] Cristea D, Vilarem G. Improving light fastness of natural dyes on cotton yarn. Dyes & Pigments. 2006;**70**:238

[186] Lee JJ, Lee HH, Eom SI, Kim JP. UV absorber aftertreatment to improve light fastness of natural dyes on protein fibers. Colouration Technology. 2001;**117**(3):134

[187] Micheal MN, El Zaher NA. Investigation into effect of UV/ozone treatments on the dyeing properties of natural dyes on natural fabrics. Colourage. 2005;**52**(Special Issue):83

[188] Gulrajani ML, Bhaumik S, Oppermann W, Hardtmann G. Dyeing of red sandalwood on wool and nylon. Indian Journal of Fibre & Textile Research. 2003;**28**(6):221

[189] Konar A, Samanta AK, Chakrabarti (Mukherjee) S, Datta S. Dyeing of jute fabric with tesu extract: Part-1: Effects of different mordants and dyeing process variables on colour yield and colour fastness properties. Indian Journal of Fibre & Textile Research. 2011;**36**:63-73

[190] Samanta AK, Agarwal P, Datta S. Studies on colour strength related parameters and compatibility for dyeing of cotton fabrics with binary mixtures of jackfruit wood and other natural dyes. Journal of Natural Fibres. 2009;**6**:27-45

[191] Samanta AK, Agarwal P, Datta S. Studies on colour interaction parameters and colour fastness properties for dyeing

of cotton fabrics with mixture of jackfruit wood and other natural dyes. Journal of Natural Fibres. 2009;**6**:171-190

[192] Singhee D, Samanta P. Studies on dyeing process variables for application of Tesu (*Butea monosperma*) as natural dye on silk fabric. Journal of Natural Fibres. 2019;**16**(8):1098-1112. DOI: 10.1080/15440478.2018.1453430

[193] IS 17084:2019. Textile Dyestuffs-Natural Indigo-Identification. New Delhi, India: BIS; 2019. pp. 1-18

[194] IS 17085-2019. Textile Dyestuffs-Rubia (Maddar)-Identification. New Delhi, India: BIS; 2019. pp. 1-14

Chapter 9

Colorimetric Analysis to Study the Dyeing Behaviour of Jute Textiles Using Selective Synthetic and Natural Dyes

Ritwik Chakraborty and Ayan Pal

Abstract

Jute, being a lignocellulosic bast fibre, is dyed with different classes of dyes such as direct, reactive, vat and sulphur which are also used to dye other cellulosic materials such as cotton and viscose. In addition to these, it is possible to dye jute with acid, basic and pre-metallized dyes due to the presence of lignin and gluconoric groups of hemicellulose. Here, a review of the dyeing of jute is done to analyse the dyeing properties of jute fabrics dyed with three classes of synthetic dyes – direct, reactive and acid dyes along with some selected anionic natural dyes. The dyeing process variables, colour strength values and colour fastness properties were analysed and elaborated. The studies have focused on some comparative assessment of the dyeing properties using different synthetic dyes and natural dyes too. The dyeing properties and colour fastness properties of synthetic-dyed and natural-dyed jute materials are also discussed and analysed in detail showing colourimetric analysis and their results.

Keywords: jute, direct dye, reactive dye, acid dye, natural dye, lignocellulosic, dyeing properties

1. Introduction

The global consciousness for sustainable ecology has gained serious momentum. Consumers want to use greener, ecologically safe products. The textile products made of natural fibres are more eco-friendly than the ones made of synthetic fibres. So, the natural fibre-based textile materials need to be produced more; similarly, they need to be discussed more. Jute among the natural fibres is different in properties and area of application. It is mostly used in packaging. No other natural fibre has more usage in packaging than jute. But, nowadays, the product range of jute is being diversified in home furnishing textiles, upholstery, automotive textiles, carpet, shopping bags, etc. These diversified products need colouration for value addition.

Jute is an important natural fibre grown in South Asia (mainly in the eastern part of India and Bangladesh) for numerous industrial applications [1]. It is the second highest plant-based fibre next to cotton grown in the world. Jute is a multicellular natural fibre, chemically lignocellulosic in nature, having about 60–64% cellulose, 20–24% hemicellulose and 12–14% lignin [2]. Numerous numbers of hydroxyl groups of cellulose and hemicelluloses, besides the glucoronic acid group of hemicelluloses having both –COOH and –CHO groups in it and several phenolic hydroxyl groups and –C=C unsaturation present in lignin, have made jute fibre highly absorptive for different classes of dyes including natural dyes with or without suitable mordants. Direct, reactive, basic and natural dyes are applied on jute. Here, the dyeing properties of various classes of synthetic and natural dyes applied on jute are discussed.

2. Dyeing of jute with direct dyes

Jute fibre, having cellulose and hemicellulose as major constituents, contains an ample amount of hydroxyl and carboxyl groups, thus making the fibre anionic while immersed in water. The anionic nature of the immersed jute fibres can be neutralised by the application of electrolyte, thus rendering the fibres conducive for dyeing with anionic direct dyes.

A few discreet studies are reported on the dyeing of jute with direct dyes [3–11]. Direct dyes are cheap but do not produce fast colour on jute fibres. Jute fibre having a negative charge in water can be dyed with direct dyes in the presence of an electrolyte which neutralises or reduces the negative potential of jute so that anionic dye molecules are absorbed on the fibre surface. So, application of an electrolyte and pH of the dyebath plays crucial roles in dye uptake of direct dyes by jute fibres. Mondal et al. studied the phenomenon and found that pH 8 is the best condition for direct dyeing of jute [4]. **Table 1** shows their results indicating the increase of dye uptake with an increase of pH till pH 8. Also, bleached jute fibres absorb more dye than raw fibres because of the presence of more free carboxyl content in bleached jute fibres.

Jute being a complex heterogeneous structure comprising hemicellulose and lignin intertwined with cellulose matrix shows its multifaceted dyeing characteristics. Ganguly et al., in an interesting study, showed the dyeing behaviour of jute with direct dyes while gradually removing hemicellulose and lignin from it [6]. The equilibrium exhaustion of direct dye and rate of exhaustion of direct dye into jute fibres rise with the gradual reduction of hemicellulose content. The authors derived three reasons for this phenomenon – (a) increase of α-cellulose concentration and primary -OH groups which are responsible for the attachment of direct dye molecules; (b) lowering of copper number, i.e., aldehyde functionality and (c) increase of fibre fineness. However, the study found out that the removal of lignin would bring about the opposite effect of direct dyeing. Dye exhaustion reduces with the reduced

	Dye uptake (%)							
	Bleached jute fibre				Raw jute fibre			
pH of dyebath →	**4**	**7**	**8**	**10**	**4**	**7**	**8**	**10**
Direct yellow 29	40.5	65.0	78.4	71.5	36.9	61.0	72.4	69.7
Direct orange 31	45.0	70.2	80.5	76.1	40.2	67.8	78.5	75.2
Titan yellow 20	20.5	32.1	46.0	37.5	17.7	28.6	37.8	35.0

Table 1.
Dye uptake of jute fibre dyed with direct dyes under different pH of dye baths.

concentration of lignin in jute fibres because of an increase in the copper number and carboxyl content in jute fibres.

The curing of dyed textile materials after dyeing has an impact of fixation of dyes into fibres. It was found that curing at 150°C gives the optimum fixation (88%) of direct dyes applied on jute-cotton union fabrics [7]. Panday et al. established a combined method of bleaching and direct dyeing of jute, thus reducing the cost and time of processing [8].

Though there are a few studies on the dyeing behaviour of direct dyes on jute, these studies were all focused on dyeing by a single dye. The behaviour of dyeing done by a mixture of direct dyes on jute was not discussed in those studies. But that is a practical aspect as industrial dyeing is generally mixing two or more dyes altogether. Such a study on dyeing of jute by direct dyes was done by Samanta et al., where six binary mixtures of six direct dyes were applied on jute and their dyeing behaviour and compatibility were studied [9]. The compatibility of dyes was analysed in the conventional method and a new method based on the colour difference index (CDI). It was found that the binary mixture of Direct Green YG (CI Direct Green 513) and Direct Turquoise Blue 2R (CI Direct Blue 199) has the highest compatibility among the pairs for dyeing of jute.

3. Dyeing of jute with acid dyes

Acid dyes generally possess sulphonic acid groups and become negative in water. This dye has relatively low molecular mass. This dye has some structural similarities with direct dye as being DSO_3Na where anionic DSO_3- generates the colour. However, they have dissimilarities – acid dyeing being nonplanar and smaller in molecular size, while direct dyes being planar and large in molecular size. This dye has been classified into three groups based on the application process:

i. Levelling acid dyes or equalising acid (low molecular weight and strongly acidic)

ii. Milling acid dyes (weakly acidic and higher molecular weight than levelling acid dyes).

iii. Super milling acid dyes (comparatively high molecular weight and neutral acid dyes).

This dye is substantive to silk, wool, and polyamide and has no substantivity to cellulose. But acid dyes have moderate to good substantivity to jute. This substantivity is attributed to the presence of nonplanar and amorphous lignin structure. However, as reported above, the study by Ganguly et al. found that the gradual removal of lignin affects the acid dye uptake adversely [6]. This is due to the increase of carboxyl content which affects the absorption of acid dyes on jute. But for hemicellulose, it is not so simple. The exhaustion of acid dye is lower than that of dewaxed jute fibre at low doses of removal of hemicelluloses, but it increases at higher removal of hemicellulose. It may be attributed to the high carboxyl content upon low removal of hemicellulose.

Though many studies on acid dyeing of jute are not available, an important study by Bhaumik et al. discussed the role of cationization of jute by the application of soya seed extract on the dyeing of acid, direct and reactive dyes [12]. As these are anionic, the dyebath needs to have an electrolyte to neutralise the negative charge generated on the jute fibre immersed in water. Cationization fulfilled the purposes and neutralised the negative charge, thus enabling salt-free dyeing of jute by anionic dyes.

Most of the acid dyes show poor wash fastness on jute, even by mild rinsing in cold water, though some acid dyes are moderately fast. But metal complex dyes have excellent wash and light fastness and therefore are most suitable for jute. Some metal complex dyes retard photofading of jute under sunlight, by UV quenching action.

4. Dyeing of jute with basic dyes

Basic dyes are cationic in nature. This class of dye is of low cost and provides brilliant shades, but it has poor to medium colour fastness to light. Many basic dyes are destroyed by chlorine, so chlorine-free water must be used in dyeing with basic dyes. This dye needs a suitable retarder for uniform dyeing for having high substantivity to acrylic or wool fibres. The retarder is not essential but is optional for dyeing jute with basic dyes. However, the metal ions present in jute fibres might change the shade of jute material dyed basic dyes, so a chelating agent like EDTA can be used to remove the metal ions present in water or in the jute itself.

A study on the optimum dyeing conditions for dyeing jute with basic dyes has been reported in the literature by Farouqui and Hossain [13].

5. Dyeing of jute with reactive dyes

Reactive dyes though being anionic in nature dye textile fibres in a way different than direct dyes. Reactive dyes react with fibres by forming covalent bonds, thus making the bond strength between the fibre and the dye stronger than the one formed in the dyeing of jute with direct dyes. So, the colour strength and colour fastness properties of the reactive-dyed jute textiles are better than jute textiles dyed with direct, acid and basic dyes.

Several studies have been conducted on reactive dyeing of jute fibres [14–20]. The effects of electrolytes on the dyeing behaviour have been examined [21]. Aluminium sulphate, alum, sodium sulphate, sodium chloride, etc. have been used for reactive dyeing of jute. Among them, aluminium sulphate is found to have yielded the highest absorption of reactive dyes by jute fibres. Alum, sodium sulphate and sodium chloride followed in succession. Similarly, aluminium sulphate achieves the highest colour fastness to light and wash for the jute material dyed by reactive dyes.

Jute can be dyed with all types of reactive dyes such as hot brand, cold brand, vinyl sulphone and high exhaustion reactive dyes. The choice of type of reactive dye depends on the shade and colour fastness needed. A study by Chattopadhyay et al. analysed the reusability of reactive dyes on jute. It was found that jute fabric can be dyed with good exhaustion if it is dyed in a separate original bath [22]. The exhaustion of dyes at the reused bath is very low for hot brand and high exhaustion reactive dyes.

When the effect of concentration of an electrolyte on the properties of the dyed jute was studied, it was found that Glauber's salt (hydrated sodium sulphate) at 25 g/L concentration achieved the best dyeing results [23].

The comparative performance of the dyeing behaviour of jute with basic dyes and reactive dyes was studied. It was found that the colour strength of the dyed jute dyed with basic dyes was higher than the one dyed with reactive dyes with the same concentrations of dyes. The reasons for this may be attributed to the higher exhaustion of basic dyes than reactive dyes. However, the colour fastness of jute dyed with basic dyes is lower than the fastness of reactive-dyed jute due to the higher bond strength of fibres and dye.

Several studies have focused on the improvement of dyeability of reactive dyeing of jute [15–18]. The application of some specific chemicals on jute before reactive dyeing has been found to be productive, enabling the fibres absorb more amount of dyes and achieve higher exhaustion with higher colour strength supported by colourimetric analysis of K/S values and ΔE values. The effect of pretreatment of reactive dyeing of jute was studied by Chattopadyay et al. [19], while the comparative assessment of different reactive dyes on jute and cotton was studied by Sinha and Roy Maulik [20].

Compatibility of soya-based natural aminio acid-modified cationic jute in between a pair of a binary mixture of reactive dyes is also reported in the literature for improving reactive dye exhaustion without salt and colour fastness to washing by Samanta et al. [16].

6. Dyeing of jute with natural dyes

Dyeing of textiles with natural dyes can be carried out in an alkaline bath, acidic bath or neutral bath. There are various reports available on different methods of mordanting and dyeing for the application of natural dyes on different fibres such as cellulosic, protein and synthetic fibres. These studies have been widely available and studied as presented in different literature. Jute is a ligno-cellulosic fibre having different functionalities, but with a high degree of crystallinity and high degree of orientation. These multiple functionalities make jute to be somewhat susceptible to dyeing with natural dyes. Still, application of mordants and modification of jute fibres are essential for better dyeing of jute fibres with most of the anionic natural dyes.

Few reports are available on chemical modifications of cotton textiles for improvement of its dyeability using synthetic and natural dyes [24–38]. However, such studies, particularly for the application of natural dyes on jute or chemically modified jute, are rare and sporadic.

There is a study [25] describing the chemical modification of the jute substrate with acrylamide monomer and post-dyeing treatment with cetyl trimethyl ammonium bromide (CTAB) for investigating their effects on dyeing properties with natural dyes such as turmeric (haldi) and madder (manjistha) substantiated with appropriate colourimetric analysis. A few studies were conducted to analyse the compatibility of dyeing of jute with various natural dyes in binary mixtures. It was found that the decreasing order of the change in chroma values of the pairs is (annatto + pomegranate) > (annatto + marigold) > (annatto + onion) and the decreasing order of the change in hue values of the pairs is (A + O) > (A + M) > (A + P). The wash fastness values for the (A + P) pair for 50:50 and 75:25 proportions are the best results [29]. The after-treatment with chitosan and CTAB improved the wash fastness by 1 grade. The application of 1% benzotriazole improved the light fastness by 1 grade for all three binary pairs. The results of the nonconventional compatibility tests were in good agreement with the results of the conventional method of plot comparisons. As per the RCR system, the compatibility grades of (A + O), (A + P) and (A + M) pairs are poor, fair and fair, respectively.

Another study [33] found that 10% harda (myrobolan) + 10% potash alum dual pre-mordanting of bleached jute fabric has the best results for colour yield (K/S values), dye uniformity (CV% of K/S values) and lower CDI (less variation or minimum dispersion of surface colour values) than that obtained by all other mordant combinations used in this study. After treatments with selective eco-safe chemical cationic dye fixing agents such as CTAB, cetrimide and also Sandofix-HCF (commercially available cationic dye fixer for comparison), it renders ½ to 1 degree of wash fastness rating due to forming salt linkages between natural dye anion in alkaline medium and cationic dye fixing agents, where CTAB is a quaternary ammonium type cation compound, while cetrimide is an amide as a potential cationiser, both of them perform equally good in improving wash fastness for all types of natural dyes applied on dual pre-mordanted jute. However, in some cases, the performance of CTAB is better than CTAB, due to differences in Pka values and cationization potential; both criteria are higher for CTAB than cetrimide. Treatment with 1% benzotriazole as an UV-absorber was reported earlier for both bleached and dyed jute cloth dyed with synthetic dyes (direct and reactive dyes) and is known to improve the light fastness by nearly one unit.

A very recent study on the statistical optimization of dyeing jute fabric with marigold petals has been reported by Chakraborty et al. [2], for obtaining more uniform dyeing on jute using alum and also by suitable double pre-mordanting system, suggesting the reaction mechanism too.

7. Improvement of colour fastness to wash and light by the application of cationic dye-fixing agents and UV-absorbers

Jute materials dyed with natural dyes tend to show poor to moderate colour fastness to wash and light, and this drawback can be corrected to some extent. Materials can be treated with several chemical auxiliaries to obtain high colour fastness to wash and light after dyeing with natural dyes. Several researchers have carried out important studies in this field.

Jute is generally treated with cetrimide, CTAB (cetyl-trimethyl ammonium bromide), for achieving increased colour fastness to wash. These chemicals make a bridge between dye molecules, thus rendering them difficult to leach out during washing. A new study showed the effect of CTAB and chitosan on jute fabrics dyed with annatto seeds. The application of 1% (owf) chitosan improved the colour fastness to washing by ½ grade and 2% CTAB improved the fastness by 1 grade. The application of 1% (owf) benzotriazole (a UV-absorber) improves the colour fastness to light of jute-dyed fabric dyed with different natural dyes by 1 grade [38].

Some chemical treatments before and after dyeing of jute with direct, acid, basic and reactive dyes are reported. Improvement of colour fastness to light of jute fabric dyed with direct dyes through after-treatment with copper sulfate and potassium dichromate is reported by Panday et al. [9], while Sarwaz et al. [11] reported the improvement in colour fastness to washing jute fabric dyed with direct dyes by treating them with various cationic dye-fixing agents. It shows that application of cationic agents after dyeing produces better fastness than the application done before dyeing.

8. Conclusions

Jute as an agro-renewable hard fibre has multiple application areas in technical textiles. Though it is still used mainly in packaging, its properties need to be studied more to bring about more diversified usages of this fibre with the help of proper scientific intervention to produce new products of woven, nonwoven, composites of jute. The fibre is abundantly cultivated in India, Bangladesh, and other South Asian countries.

The present article has shed significant light on the dyeing behaviour of jute and its bottlenecks. Colouration of jute textiles is an important area of research and development for the manufacture of potential end-products in global markets. Dyeing of jute with direct dyes and basic dyes is economic and possesses a wide range of shades, but this suffers from having low colour fastness to wash and light. Some works have been narrated here to overcome the obstacles of these dyes. Reactive dyes being colour-fast and possessing good shades are costlier than direct, acid and basic dyes.

As jute is an eco-friendly fibre, this special aspect of the fibre can only be enhanced if it is dyed with eco-friendly natural dyes. Here, dyeing of jute with natural dyes has been discussed, and suitable after-treatment processes to overcome the low colour fastness associated with natural dyeing have been mentioned.

Author details

Ritwik Chakraborty* and Ayan Pal
Department of Jute and Fibre Technology, University of Calcutta, Ballygunge Campus, Kolkata, West Bengal, India

*Address all correspondence to: ritwik.chak@gmail.com

References

[1] Samanta AK, Mukhopadhyay A, Ghosh SK. Processing of jute fibres and its applications. In: Kowzłowski R, Mackiewicz-Talarczyk M, editors. Handbook of Natural Fibres: Processing and Applications. Vol. 2. Cambridge, United Kingdom: Woodhead Publishing; 2020

[2] Chakraborty R, Samanta AK, Vankar PS. Statistical optimization of process variables for the dyeing of jute with marigold petals using a dual mordant system. American Association of Textile Chemists and Colorists Journal of Research. 2024;**11**(1):20-36

[3] Samanta AK, Das D. Studies on quantitative colour measurement of direct dyed jute fabric in relation to computerized colour matching. Journal of Institution of Engineers (India): Series E. 1992;**73**:53-57

[4] Mondal MIH, Islam MK. Effect of pH on the dye absorption of jute fibre dyed with direct dyes. Oriental Journal of Chemistry. 2014;**30**(4):1571-1575

[5] Farouqui FI, Hossain I. Dyeing of jute fabric with direct dyes and their fastness characteristics. Indian Journal of Fibre & Textile Research. 1990;**15**:65-72

[6] Ganguly PK, Chanda S. Dyeing of jute: Effect of progressive removal of hemicellulose and lignin. Indian Journal of Fibre & Textile Research. 1994;**19**(1):38-41

[7] Chowdhury N, Iqbal SMF, Saifullah K, Chowdhury TA, Khatun M. Study on continuous dyeing of jute-cotton union fabric with direct dye. Saudi Journal of Engineering & Technology. 2021;**6**(6):139-142

[8] Panday SN, Chattopadhyay SN, Day A, Pan NC. Single bath bleaching-dyeing of jute fabric with hydrogen peroxide-direct dye combination. Indian Journal of Fibre & Textile Research. 1994;**19**(1):34-37

[9] Panday SN, Day A, Chattopadhyay SN, Pan NC, Mathew MD. Effects of after treatments on direct dyed jute fabrics. Indian Journal of Fibre & Textile Research. 1993;**18**(3):87-90

[10] Samanta AK, Bhaumik N, Konar A, Roy AN. Studies on compatibility of selective direct dyes for dyeing of jute fabric. Indian Journal of Fibre & Textile Research. 2019;**44**(1):98-106

[11] Sarwar Z, Azeem A, Munir U, Abid S. Direct dyeing of jute: Effect of cationic treatments on color fastness. Journal of Textile Science & Engineering. 2017;**7**(2):1-4

[12] Bhaumik N, Konar A, Roy AN, Samanta AK. Comparative studies on dyeability with direct, acid and reactive dyes after chemical modification of jute with mixed amino acids obtained from extract of waste soya bean seeds. Journal of Institution of Engineers (India): Series E. 2017;**98**(2):121-133

[13] Hossain I, Farouqui FI. Study on the optimum dyeing conditions in dyeing jute with basic dyes and their fastness under various influences. Textile Dyer & Printer. 1990;**23**:28-32

[14] Samanta AK, Chakarbarti Mukherjee S, Guharoy TK. Dyeing of jute fibre with reactive dyes: Optimization of process variables and assessment of colour fastness characteristics. Journal of

Institution of Engineers (India): Series E. 2012;**93**(1):15-24

[15] Samanta AK, Konar A, Kar TR, Chakraborty S. Effects of amino-acid modification of jute fabric on its textile related properties and dyeability with selective reactive dyes. International Review of Applied Engineering Research. 2014;**4**(1):51-67

[16] Samanta AK, Bhaumik NS, Konar A, Shome D, Roy AN. Studies on reactive dyeing of natural amino acid modified cationized jute fabric: Colour interaction parameters and colour fastness properties. Journal of Materials, ASSCIT, USA. 2015;**1**(4):89-97

[17] Samanta AK, Kar TR, Mukhopadhyay A, Konar A, Shome D. Studies on dyeing process variables for salt free reactive dyeing of glycine (amino acid) modified cationized cotton muslin fabric. Journal of Institution of Engineers (India)-Series-E. 2015;**96**(1):31-44

[18] Samanta AK, Konar A, Mukherjee S. Salt-free reactive dyeing of amino-acid modified jute. Indian Journal of Natural Fibre (TINFS). 2014;**1**(1):15-28

[19] Chattapadhyay SN, Pan NC, Day A. Ambient temperature bleaching and reactive dyeing of jute: The effects of pretreatment, bleaching and dyeing methods. Journal Textile Institute. 2002;**93**:306-315

[20] Sinha DK, Roy Maulik S. Comparative studies on dyeing of jute and cotton with different classes of reactive dyestuffs. Textile Trend. 2002;**45**(7):25

[21] Salam MA, Sheik RK, Farouqui FI. Effect of salts on dyeing into jute with reactive, direct, basic and mordant dyes. Journal of Textile and Apparel Technology & Management. 2009;**6**(2):1-6

[22] Chattopadhyay SN, Pan NC, Day A. Reuse of reactive dyes for dyeing of jute fabric. Bioresource Technology. 2006;**97**:77-83

[23] Rahman MM, Biswas J. Effects of salt in dyeing of jute fabric with reactive and basic dye. International Journal of Latest Trends in Engineering and Technology. 2019;**13**(3):39-43

[24] Samanta AK, Singhee D, Sengupta A, Rahim SK. Application of selective natural dyes on jute and cotton by different techniques. Journal of Institution of Engineers (India): Series E. 2003;**83**:22-33

[25] Samanta AK, Agarwal P, Datta S. Physico-chemical studies on dyeing of jute textiles with natural dye extracted from red sandal wood. Journal of Institution of Engineers (India): Series-E. 2006;**87**:16-26

[26] Samanta AK, Agarwal P, Datta S. Dyeing of jute and cotton fabrics using jackfruit wood extract: Part 1: Effects of mordanting and dyeing process variables on colour yield and colour fastness properties. Indian Journal of Fibre & Textile Research. 2007;**32**:466-476

[27] Samanta AK, Agarwal P, Datta S. Physico-chemical studies on dyeing of jute and cotton fabrics using jackfruit wood extract as natural dye: Part II: Dyeing kinetics and thermodynamic studies. Indian Journal of Fibre & Textile Research. 2008;**32**:66-72

[28] Samanta AK, Agarwal P. Application of Mixtures of Red Sandal Wood and Other Natural Dyes for Dyeing Jute Fabric - Studies on Dye Compatibility. International Dyers; 2008. pp. 37-42

[29] Chakraborty R, Samanta AK, Vankar PS. Studies of compatibility of dyeing of jute with binary mixture of natural dyes. International Journal of Innovative Research in Science Engineering & Technology. 2022;**11**(11):13424-13429

[30] Chakraborty R, Samanta AK, Vankar PS, Samanta P. Review on extraction, Mordanting and application of natural dyes on jute. Current Trends in Fashion Technology & Textile Engineering. 2022;7(4):1-10

[31] Samanta AK, Agarwal P, Datta S. Dyeing of jute with binary mixture of jackfruit wood and other natural dyes - Studies on colour performance and dye compatibility. Indian Journal Fibre & Textile Research. 2008;**33**:171-180

[32] Samanta AK, Agarwal P. Review on application of natural dyes on textile materials. Indian Journal Fibre & Textile Research. 2009;**34**(4):384-399

[33] Samanta AK, Singhee D, Agarwal P, Datta S. Application of single and mixture of red sandal wood and other natural dyes for dyeing of jute fabric: Studies on colour parameters/colour fastness and compatibility. Journal Textile Institute. 2009;**100**:565-587

[34] Samanta AK, Agarwal P, Datta S. Studies on colour strength related parameters and compatibility for dyeing of cotton fabrics with binary mixtures of jackfruit wood and other natural dyes. Journal of Natural Fibres. 2009;**6**:27-45

[35] Samanta AK, Agarwal P, Datta S. Studies on colour interaction parameters and colour fastness properties for dyeing of cotton fabrics with mixture of jackfruit wood and other natural dyes. Journal of Natural Fibres. 2009;**6**:171-190

[36] Samanta AK, Agarwal P, Konar A, Datta S. Characterisation and application of purified selective natural dyes on jute. International Dyer. 2008;**193**(3):25-33

[37] Chakraborty R, Samanta AK, Vankar PS. Dyeing of jute with annatto seeds and other natural dyes – Studies of compatibility and other parameters. Indian Journal of Fibre and Textiles Research. 2023

[38] Chakraborty R, Samanta AK, Vankar PS. Functionalization of jute to improve colour yield and fastness for dyeing with annatto seeds. In: Proceedings of Third International Conference on Functional Textiles and Clothing 2023. USA: Springer Publisher; 2023